Discovering the New View of Safety

DISCOVERING THE NEW VIEW OF SAFETY

USING LEARNING TEAMS

Moni Hogg

Published by Moni Hogg

First published in 2024 in Auckland, New Zealand

Copyright © Moni Hogg

www.monihogg.com

Edited by Jenny Magee

Designed by Monster Illustration & Design

Typeset in Australia by BookPOD

ISBN: 978-0-473-71748-3 (paperback)

ISBN: 978-0-473-71749-0 (ebook)

Haere taka mua, taka muri; kaua e whai.
Be a leader, not a follower.
(Māori proverb)

Acknowledgements

Books are rarely written in isolation, and this book has certainly benefitted from so many people who contributed generously through collaboration, conversations and case studies. I am grateful for their support for the development of my thinking.

The following stand out as amazing colleagues doing fantastic work: Briar Moffatt, Chris Eastham, Matt Ward, Tony Mills, Tiana Epati, Rebeca Clifton, Keith Bardwell, Brian Shea, Hema Puthran, Daniel O'Connor, Stephanie Weal, Peter Murran, Graham Bradley, Paul Robertson, Steve Slothouwer, Mark Green, Brian Cleur, Bruce Kemp, Hans Key, Tiketi Auega, Rachael Trimble, Angelique Fraser, Richard Burke, Katie Hunter, Dale Ewers, Richmond Johnston and Jason Hare. Please forgive me if I've missed anyone.

Firstly, I'd like to acknowledge Tristan Casey, a highly respected international academic and consultant who truly embodies the values of our movement. Thanks so much Tristan, for your ever so generous support with structure and peer review of material. You are a rockstar!

John Skudder has been a great friend on this journey. He is kind and generous to all and a mentor to many. Thank you for your unwavering support over recent years.

Bruce McDonald, Patrick Seaman, and John Robertson were wonderful mentors in my formative years. Each died far too young (in their 60s), but I've never forgotten their wise guidance.

There's a phenomenal team collaborating across New Zealand to handhold Safety Differently through. Over recent years, our safety profession has become an awesome tribe – connecting, sharing ideas and making things happen. It's enormous fun and very exciting to see so much change in such a short space of time. My special thanks go to Francois Barton, Jon Harper-Slade and Selena Armstrong for their kindness and encouragement in recent years.

I also want to acknowledge those New View Safety thinkers who have influenced and informed my work. As the adage goes, no ideas are new; however, the creativity, passion and time to lead and support a new movement require more determination, grit and courage than others might see.

Daniel Hummerdal is a great inspiration, and his leadership in the New Zealand safety sector has brought immeasurable change. He has deftly created a space for improved ethics to emerge to full maturity.

Sidney Dekker's leadership from a place of ethics and values is changing the world, and I salute his example.

Drew Rae's early encouragement and time were invaluable. He continues to challenge the status quo and is globally respected.

Thanks to Kelvin Genn and the Art of Work team, past and present. Kelvin is a genius thought leader whose concepts and critical thinking have strongly influenced me.

Todd Conklin developed the Learning Teams principles drawn upon in several chapters in this book. I acknowledge the valuable contribution Todd continues to make to our sector.

Finally, and most importantly, I want to acknowledge all who have bravely taken the innovation journey in recent years. The path to success is littered with the pain of failing forward. Every safety lead, every organisation and every thought leader who has experienced such failure over the past decade has contributed to the success of the New View Safety movement. Many people worldwide have played their part. It's when we all go forward together that we achieve great things.

Contents

Introduction   1

1: Why Shift from Current Safety Approaches   9

2: Introducing New View Safety   35

3: The Readiness Test   57

4: Foundations of Success   69

5: Taking the Change Management Journey   81

6: Introducing New View Ideas to Your Leadership   97

7: Introduction to Learning Teams   119

8: Strategies for Facilitation   141

9: Learning Reviews   165

10: Contractor Partnerships   179

11: Your Agile Learning Team Framework   187

12: Improving Safety Reporting and Assurance   209

Epilogue   223

About the Author   233

References   235

Introduction

Welcome to the exciting world of New View Safety, which radically redefines how we create safety and wellbeing in our organisations. Traversing this territory is often confusing and paradoxical, yet it is enlightening and requires courage and bravery.

The purpose of this book is to set you and your organisation on the journey to New View Safety. If you're already on your way, use this book as a refresher and an opportunity to refine what you're doing. I hope the ideas and practical suggestions will inspire you to go further.

This book, *Discovering the New View of Safety,* maps the move from current safety systems to new philosophies and methodologies, crossing the landscape of Learning Teams.

My next book, *Evolving the New View of Safety*, is already in draft. From the foundation of Learning Teams, it moves safety practices towards the principles found in self-directed teams.

Together, these two books present wisdom from the emerging field of safety science based on my personal practical implementation and the experiences of others. Translating theory into practice is the current challenge.

A couple of quick notes.

New View Safety is an umbrella term for contemporary safety science that is gaining ground internationally. In the interests of variety, I'll circle through synonymous terms, including Safety Differently and Safety II. More definitions later.

In a world where technology is at our fingertips, I've been asked whether I employed AI tools in writing this book. While those tools have value in some settings, I have not used them here. For better or worse, these words are my own.

Why this book?

Our current societies have been built over millennia. Whatever your political bent, it's not hard to see that we have created structures and systems that attempt to institutionalise morality.

Where religion was once the predominant mechanism for keeping basic social ethics intact, we now have dedicated laws, governments and agencies. The safety science profession has arisen from the increasing demand for ethics in business following the industrial revolution.

Society, however, is going through extraordinary new shifts. These are, in part, driven by the exponential growth in technology, which has democratised knowledge and is increasingly democratising power.

Technology has created a new form of global connectedness. Yet our institutions, structures and systems are creaking and groaning. In all walks of life, we seek new solutions to the age-old question of ethical leadership.

> Our organisations desperately need changing. They are rife with parental attitudes towards employees, politics, infighting, bureaucracy, bullying, poor wellbeing, and overall discontent. All of which are ultimately driven by fear. This negativity affects large parts of our population.

While we often struggle to see a more harmonious way of working and living together, there is a growing call for a shift in consciousness. From blame to trust. From control to transparency. From protection to connection. From centralised power to self-determination.

Regardless of personal opinions, how can the safety world contribute to the wider worries facing us today? How can our little corner of the world make a difference in the bigger scheme of things?

Every organisation is on its own journey of evolution; yours will be no different. Your amazing teams and leadership, and the customers, communities and nations you serve are vital parts of our planet's complex web of life.

We are not alone

Safety thinking is currently undergoing a major shift, although, at its core, it's largely a shift in focus. Similar transitions are happening globally in other fields, so we're not alone in a fundamental change in methodology.

Figure 1: A massive shift in worldview

By the fifth century BC, it was generally accepted that the earth was round rather than flat. This was a game-changer and had a major influence on the prevailing paradigms.

Equally significant changes are occurring in science now.

Corporates, who arguably run much of the world, currently focus on avoiding error, driven by the scientific management approach of the twentieth century. This has resulted in the rules, procedures, bureaucracy and control that have become synonymous with organisational safety. It has created a behemoth that is out of control.

> In the early twentieth century, scientists made radical discoveries known as quantum physics. We now recognise that focus (the influence of the observer) impacts behaviour and outcomes. Neuroscience shows that our brains develop according to what we pay attention to.

These scientific understandings are filtering into applied science and business methodology. For example, the four-day workweek movement focuses on output rather than time. In animal welfare agencies, the shift is from preventing harm to improving positive experiences for animals.

In New View Safety, the move is from avoiding error to building team capacity and resilience. As with motorsport, we are focused on getting into the gap. Our sector's journey over the coming decade and beyond is to fully understand what that gap is. We've got a lot to discover.

New View Safety detractors seem to think that tried-and-true wisdom is being thrown out with this shift. It's not. All the good safety practice we've developed over time is needed now as much as ever. What we're doing is evolving – nudging our systems forward based on new thinking and a new focus. We're taking the good and aiming to make it great.

So, what do we want to focus on then? We know that rules, procedures and punishment have only gotten us so far. What, then will get us to the next stage of evolution? We need a radical new way of looking at how we work together.

There's a lot to this picture of a brighter future; however, three keys involve building trust, transparency and honesty throughout our social structures. Only then can we reduce the excessive need for regulation, control and bureaucracy. It's a new landscape to discover. And innovation is the vehicle.

A call to action

Supporting the New View Safety movement and being a change agent in your corner means observing better ethics and humanising your systems. That leads to improved performance, ways of working together, and (hopefully) better structures. You become part of the vision of a society with a better future.

I've written this book (and the next) because, over my twenty or so years as a safety professional, I've consistently seen great safety leadership expressed by individuals and teams that hierarchical organisations don't fully harness.

No wonder others are frustrated with our profession. The primary reason is that most organisations are overly focused on the wrong things. They are blind to the true potential of teams to create the solutions they seek. Their systems and structures are holding them back.

Our people across industries, nations and around the globe are truly remarkable. I want to encourage my colleagues everywhere to engage in experiments and pilots to redesign how safety leadership is structured.

I'm not talking about slicker, fancier initiatives based on the benefits of caring about each other (although these have their place). I'm talking about changing systems and structures to harness our full potential.

This call to action is in the context of current wider social change. The Millennial and Gen Z generations appear to embrace a progressive culture with a new set of values. They intrinsically understand that everyone has value and that mutual support is essential to achieving potential.

> The competitive hierarchical structures that new generations have inherited don't fit their ethos and will crumble once they lead and govern our organisations. Therefore, those of us with the power to make change now must prepare the way.

While New View Safety is backed by recent advancements in safety and social science, my attempt with this book is to give a practitioner-based interpretation of the ideas.

I share this material on the basis of 'take what you like and leave the rest'. At the same time, I encourage every safety professional to boldly and courageously further the collective innovation journey from your corner.

Take the ideas in this book that resonate with you, and experiment with them. Share and publish the results. Be part of the change and enjoy it.

My request

Please pop me a note on LinkedIn once you've read the book – I love making connections around the world. Enjoy.

CHAPTER 1

Why Shift from Current Safety Approaches

Current safety methods have gotten us so far – it's time to evolve

Key points

- The Industrial Revolution massively improved productivity but dumbed down workers.
- Management thinking has created a false construct that workers are less capable and need to be controlled.
- Contemporary safety thinking creates transparency over system vulnerabilities and sees that capable teams are successful despite imperfect conditions. The approaches meet the intended requirements of directors in our safety legislation.

'Our HR manager has taken some of the recommendations out of the Learning Team report,' he said with a resigned sigh. My client, the safety lead for the small Company Z, had been facilitating Learning Teams – an enquiry-based workshop with frontline workers – for a couple of years. He supported the workers voicing their concern about the culture around housekeeping and cleanliness in their processing plant.

The work teams had noted that leadership were either turning a blind eye or were unaware of the issue because they weren't resourcing the teams particularly well to handle it themselves. Production output was paramount. Rodents were starting to appear in the facility, alongside the prevalence of dust which made surfaces slippery.

The workers had made several suggestions for how to improve the situation. Their preferred solution was support from leadership via walkarounds and generally taking an interest in the situation. The HR manager removed this recommendation from the report.

We will assume the HR manager (a good person who cared a lot for the team) had best intentions. They would have thought they were doing the right thing. Almost all individuals strive to be ethical in their dealings; so we must look to the systems and structures when seeking change.

The question, though, is why this would happen when the leaders and directors should (and have a legal duty to know) about such a situation. We'll return to Company Z later in the chapter; for now, we need to understand how an HR manager can have such authority over team insights and what that means for our organisations.

History of labour relations and management philosophy

For much of our history, humans lived in nomadic tribes and bands. It wasn't until the Agricultural Revolution and the notion of land ownership came about that organisational structures developed based on competition.

Before the Industrial Revolution, most work was done through the craftsmanship of butchers, bakers, builders and others. Women were cooks, made clothing and bought up children. Work was specialised based on trades, and the responsibility for workmanship rested squarely with the individuals carrying out the work.

The Industrial Revolution of the twentieth century changed all that. Taylorism (aptly coined scientific management) and the associated specialisation of labour, stripped workers of opportunities to participate in planning, evaluating and improving work processes. The responsibility for the quality of the work now rested with a new class – the managers.

In today's organisations, we see top-down decision-making, prescribed tasks and narrow job descriptions. Industrial relations, work rules and personnel policies prevail. Bosses control a system designed to limit employees' room for error.

This obsession with hunting out and controlling error has created the quality and proliferation of products and services we enjoy today in an increasingly technology-driven world. No wonder we are resistant to change. Who doesn't love the relative comfort of our lives compared to a hundred years ago?

However, there is growing discontent with the dehumanising ways of organisations. Science shows that the focus of mainstream organisations on 'predict and control' (a desire to predict the future and control all or most outcomes) needs an upgrade. More on this soon.

Rob Long discusses this dehumanising effect in his article, *The Ethics of Safety*.[1] He says:

> 'The key to the effectiveness in safety is the humanisation of others and the building of relationships. It is strange that the vice of intolerance is advocated in the workplace, but we would never want such a vice in our home relationships. The challenge for the safety community is to understand the impact of negative ethics and seek positive ethics in the promotion of care, safety and the management of risk.'

Sidney Dekker and the Safety Differently community have declared that safety should be framed as an ethical responsibility rather than a constant bureaucratic focus. An international safety science community posited a new framing of safety to humanise workers and answer the question of ensuring moral outcomes in society.

Off the back of this, the New View Safety movement has burgeoned. The great news is that these approaches simultaneously improve ethics and performance. Just ask any of the organisations who are early adopters and have made sincere attempts to apply the new theory. They will attest to what I'm saying. The win-win outcomes are good for the business, which we'll see throughout this book.

The vision of the movement is, over time, to reduce reliance on regulation while driving the desired social outcomes. Less regulation and bureaucracy is good for business. If business takes full responsibility for ethical outcomes, compliance, which is inefficient, can be reduced. It becomes an upward spiral. Who's going to argue with that?

Before we start on how to do all this amazing stuff, let's look closer at the problems created by how we currently work together.

Workers have become dependent on managers and systems

While we rarely question the hierarchical nature of most of our organisations and institutions, hierarchy creates the need to please managers, which drives conformity.

This management system (not the only available option despite its prevalence) creates employees driven by management rather than customers and conformity rather than creativity. Managers are rewarded with status and power, which can conflict with good leadership.

This system essentially dumbs down most people who must conduct work based on predefined role descriptions designed around predict and control thinking. It performs poorly at harnessing the true potential of those tasked with bringing forth the purpose of an organisation.

Ironically, we spend much time and money trying to reverse these effects with leadership training, innovation training, staff development and more. Yet, we don't ask whether the structures serve the purpose for which they are designed.

Organisations are stifled with endless meetings, politics, infighting and bureaucracy. Blame, finger-pointing, low trust and poor relationships prevail. Performance management is seen as the answer, and discipline deals with accountability. A parent-child construct for directing work and solving problems emerges.

A common symptom of this is teams bitching and moaning about problems in the organisation, constantly frustrated that their feedback and ideas go unheard. Overly busy managers try to have a handle on too many things and are frustrated when workers don't take more responsibility for resolving issues.

Organisations become limited by the capacity of their leadership.

We need, instead, to build structures that activate the 'sense and respond' process (more on this soon), which enables workers who are close to the work and understand what's needed, to do something about it.

Our current ways of working often shut this down, reducing creativity and agility and creating situations where staff pay attention to what keeps their boss happy rather than the client. You've likely seen this: staff with no authority to make decisions end up following dumb processes to avoid getting into trouble, instead of doing what would be in everyone's best interests.

With a traditional boss/worker relationship, the locus of innovation is always top down. Teams don't offer their creative ideas anywhere near the extent to which the organisation could benefit. Or, when they do, follow-up and implementation are poor, and the teams decide it's a waste of time.

We need a shift towards redesigning structures to build the capacity of teams to solve these problems. This approach is *far* more powerful as it generates capacity to handle the variability of demands that are a natural part of our work. (The safety science community terms this ability 'resilience', and it is the backbone of the school of thought called resilience engineering.) Pioneers have shown the way, and we will draw from these throughout this book to understand how we can adapt and evolve our safety programmes.

With organisations intent on building agile and creative teams, including improving safety outcomes, the foundation is unleashing potential. As mentioned earlier, the most often reported challenge is the transition needed in managerial thinking. New skills, perspectives and strategies are essential, although some may perceive them as threatening their power and status.

In addition, workers are used to their dependence on bosses and may need support to participate further. They often need to develop the confidence that they will be heard and supported to make mistakes as they learn and grow. Participative management accepts that teams can build the capacity to manage and lead themselves. They can furnish the initiative, sense of responsibility, creativity and problem-solving from within. In short, teams are capable of being self-reliant.

From 'predict and control' to 'sense and respond'

Given the problems organisations face, the challenge is understanding the evolution stage required to address these needs. If that seems like a task beyond your organisation's ability, the missing epiphany is that you must design iterative steps to build team capacity and resilience.

Before we unpack the steps and stages of the journey, we need to understand what we're trying to achieve. We need to enable teams to 'sense and respond'. This term is referred to in complexity theory, which underpins New View Safety. First, however, let's talk about it in practical terms.

As they go about their work days, teams sense when work isn't flowing right. They connect to issues that arise and see better ways of doing things. Each team member has a unique perspective and a window into how things could be improved. Teams and departments have their own perspectives as well.

The hierarchical, controlled, top-down nature of work means these individuals and teams often can't respond (i.e., change and make improvements) without getting busy managers on board or dealing with politics and bureaucracy. Organisational design is usually very frustrating in this regard. We all have war stories.

Organisations are complex adaptive systems, meaning they adapt and change organically rather than according to some prescribed plan. Approaches that harness the intelligence of their people massively improve creativity and innovation and achieve the purpose of the organisation.

Teams are amazing and can successfully handle the difficulties surrounding hierarchical organisations. They make do with overly prescribed procedures, imperfect resourcing, challenges in resolving goal conflicts and a raft of other issues. In Safety II language, they are highly adaptable actors who create success in almost all conditions and despite the constraints, day in and day out.

Improving ethics by humanising people supports better performance in our businesses. It unleashes creativity, efficient problem-solving and the ability of the teams to make effective decisions that they own and take responsibility for. This is the promise of Safety II and it's truly a win-win.

Using the principles of self-organisation to build capacity and resilience

I've had the privilege of setting up (almost from scratch) a framework based on Safety Differently principles for a high-tech organisation in a high-risk environment with critical safety needs. The CEO wanted safety management that was responsive to a fast-paced, rapidly changing environment with a team of capable and engaged staff.

A traditional command and control style of safety didn't suit the needs of the organisation or the people. Creating bureaucracy would only hamper the organisation's need to remain as agile as possible.

A safety department could provide well-meaning advice on the systems required to enable safety in a reasonably stable environment; however, the design engineers and technicians were by far the most qualified to make on-the-ground decisions.

At that time, with fresh ideas coming from academics and international leaders such as Sidney Dekker, the organisation decided to try a bottom-up approach where teams had full autonomy to manage their own safety, create rules and procedures according to their needs, and take full responsibility for identifying all risks and finding the solutions and controls for them.

The result was shared safety leadership amongst teams with high levels of participation, collaboration and conversations about safety. Unnecessary bureaucracy was reduced to a bare minimum to meet only specifically prescribed regulations. Teams had the flexibility to be responsive to the agile, innovative and rapidly changing working environment. This was enabled via a high trust, blame-free culture, an authentic open-door policy with senior leadership and radical transparency with regulators.

Teams were empowered to implement controls and change rules about their safety according to current conditions. Clear backing by the CEO and leadership team meant adequate resources were available for the teams. Engagement was extremely high, with working groups formed around key risks, safety training and procurement.

While all this may seem impossible for your organisation, you are already using some of the principles of self-organisation. At its core, self-management means knowing exactly what you are responsible for and having the freedom to meet those expectations however you think is best. Self-organisation is being able to make changes to improve things – beyond what is required of you. Essentially, all that is needed is commitment from everyone to make it work!

The best way to enable teams to 'sense and respond' is to set them up for successful self-organisation to whatever extent is possible. That means those doing the work have the best understanding of the risk involved and are, therefore, better placed to make the best decisions on handling that risk. Obviously they'll need some support, but less than our organisations fearfully think.

To self-organise, teams need the following:

- authority to make decisions
- resourcing to ensure the necessary capacity and capability
- transparency of information to support decision-making
- a means to support peer-to-peer accountability.

As the New View concepts have been accepted and trialled in recent years, we have seen aspects of these principles used with new tools and methodologies, including the Human Organisational Performance (HOP) principles, Learning Teams, and tools based on resilience mapping.

However, the aim over time is to build these add-on approaches into the main fabric of your organisation so that managing safety is not separate from how the rest of the organisation functions. While all this may seem too big a step change right now, you need to consider how your roadmap could look over the next decade and beyond.

As the adage goes, don't overestimate what you can achieve in one year and don't underestimate what you can achieve in ten. For now, see this as a line of sight to where you are heading and aim to break down the steps to get there. With the level of risk and the regulatory frameworks they must adhere to, increasing trust in front-line teams is a challenge for some organisations.

They worry about legal defensibility. They worry about the capability of the teams to step up. They worry about the robustness of controls and whether they can still be adequately verified. There are answers to all these concerns, as we'll see. Plenty of new methodologies and tools are available to nudge us in the right direction.

Over time, you'll need to tackle these issues using innovation techniques and new methods to progressively build capability in your organisation. This has to be done intentionally and requires forethought, commitment and courage to lead the necessary change. We'll unpack all this across successive chapters plus the next book.

For now, I can confidently share that five years on from implementing Safety Differently, the high-tech organisation just mentioned has gone from strength to strength in performance with minimal, if any, injuries to the team. A contemporary safety programme has been one factor in their overall success.

The roadmap from here

The Safety Differently journey is about building trust and ownership to strengthen the adaptability and resilience of your teams. When you start with easy initiatives, this gains momentum and more becomes possible. It means growing the capability of teams that have become dependent on the systems they work within.

The other challenge is shifting management to a different way of working with their teams. Once you get going, you'll see the potential of this approach and only want more. Naturally, you will need to maintain appropriate levels of control while testing ideas, adapting and evolving.

Some leaders choose not to put time and resources into New View Safety initiatives because they have other priorities, but this is a mistake. The very problems they are tied up solving will be readily answered and resolved by these new approaches.

If you take small steps, it's not hard to nudge your system with well-designed experiments and small initiatives and use them powerfully to demonstrate new ideas to your organisation.

It is then up to everyone to adopt the ideas and drive them organically through the system. Your job is to catalyse, support and facilitate this process. Operate as a change agent. We'll explore this in detail soon.

You'll have plenty of teething problems along the way. The good news is that you can go slow, steady and under the radar if necessary. Creativity and commitment are the essential ingredients.

Stages of empowerment

Most safety programmes still lean on committees, surveys and other accepted instruments to deliver worker participation and engagement. They are accepted means of discharging legal duty and are written deeply into our auditing and assurance programmes. These mechanisms have driven improvement in recent decades and prevail despite better alternatives. But the results are plateauing and, as Figure 2 shows, they are at the lower end of the continuum of empowerment.

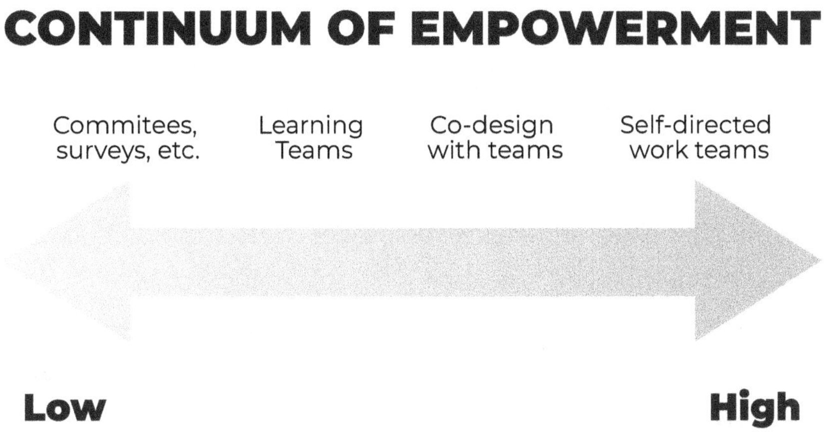

CONTINUUM OF EMPOWERMENT

| Commitees, surveys, etc. | Learning Teams | Co-design with teams | Self-directed work teams |

Low **High**

Figure 2: Continuum of empowerment

Learning Teams are a structured process designed to facilitate team-based enquiry about everyday work from the perspective of the team members. We empower them by asking what improvements would make it easier to adapt successfully to challenging day-to-day work conditions.

Through a structured, facilitated exercise, we enable teams to share what they 'sense' in conversations with their managers, who, hopefully, support their insights by making the necessary improvements (providing the 'response'). In this way, Learning Teams are a discrete, micro, self-directed exercise.

This process begins the shift to self-directed work teams as we teach management to listen to the reality of how the work is done and co-design improvements with the team.

Learning Teams build trust between workers and managers, which is the crucial ingredient for shifting to higher levels of empowerment. My next book will further explore the components that are necessary to embed trust further and give more authority to the teams.

Learning Teams and other work insight methods are becoming accepted across Australasia. In upcoming chapters, we'll detail how to do the Learning Teams process because they're important stepping stones as you build your New View Safety framework. They start the necessary journey of building transparency up and down the organisation – between frontline teams and the board directors who prioritise resourcing. The upward spiral of trust starts here.

I accept that there are challenges with Learning Teams. Common concerns expressed include the time-consuming nature of the sessions, inadequate follow up, and that it often takes champions to keep the methodology alive in the business. However, they are definitely worth the time because workers' insights can usually fix previously unsolved problems. Win-win outcomes mean managers realise the value immediately.

One high-risk company was the pioneer for Learning Teams in New Zealand. Champions, including operational leads with a passion for the new work and safety lens, drove change through the business. This led to a genuine shift from blaming workers when something went wrong, to learning along with them.

Ten years on, they are still as passionate about Learning Teams. Operational teams request the process and ask for more facilitators to be trained when some leave the business. Everyone understands and recognises the ethos. Known as a high-performing organisation, there's no doubt that the Learning Teams programme is part of their success. At the time of writing, they are currently looking at extending their programme to include more advanced ways of empowering teams by driving resource and budget prioritisation as close to the team level as possible.

Stages of the journey

This book takes you through the initial stages towards an entire system based on the Safety Differently principles, focusing on the first three stages of readiness, starting out and having a structured framework as shown in the following table.

My next book, *Evolving the New View of Safety*, will complete the journey, describing the remaining elements of co-designing systems, ethical leadership and self-directed work teams. All the elements required to evolve your entire system. While that may seem unrealistic right now, remember how fast society is changing. Buckle up; we're in for a ride.

For now, though, our focus is on the initial three stages, set out in Table 1.

STAGE	PURPOSE	ACTIVITIES	OUTCOME
Readiness	Introducing concepts	Trials and experiments	Understanding
Starting out	Demonstrating the change in focus	Human organisational performance (HOP) principles	Trust
Structured framework	Embedding the philosophy	Learning Teams, worker insights capture and reporting	Listening
Co-design systems with your teams	Building team capability	Integrate safety management with work	Ownership
Ethical leadership	Legitimately and safely reducing bureaucracy	Transparency with regulators, peer-to-peer accountability	Connection
Self-directed work teams	Enable and resource teams to self-determine most of their safety needs	Team-based safety systems with the authority to adapt autonomously	Performance

The first three stages are grouped under **Discovering the New View of Safety**, and the last three under **Evolving the New View of Safety**.

Table 1: The roadmap to implementing Safety Differently

The HR manager at Company Z had good reasons for editing the report. In this case, they were trying to protect senior colleagues (who were busy resolving strategic matters for the business) from operational details. However, despite their good intent, the law says that directors and those responsible for carrying out their duties must understand what's going on at a team level and resource those needs accordingly. This team was alerting their seniors to a genuine issue that needed visibility to be resolved. Their insight should have been trusted and transparency enabled.

Directors have a legal and moral duty to know, understand and resource the risks. In practice, that means recognising the insights from the work teams who fully know what's going on and ensuring they can effectively resolve problems. Managers, the safety team and other experts have a limited picture created by the hierarchical nature of organisations. We need a New View of how to deal with this. In addition, work teams need to understand the director's strategy and decision-making, so they can adapt effectively, and know what intel needs to flow back upwards.

Throughout this book, we'll explore the early territory that is well underway, culminating in Chapter Twelve, which details how to turn the new methods into a measurement system (with a view to using AI to handle the data) that achieves transparency over what's happening with teams. To do that, we need to re-examine how hierarchical relationships affect how we work together and start iteratively making change.

The next book, *Evolving the New View of Safety*, will dive into relatively uncharted territory. We'll talk about how to iteratively build ownership and capability so teams can safely adapt to daily conditions in an agile manner with authority, resourcing and collective decision-making. This requires integrating safety systems with operational work (instead of a siloed system). It takes a fully decluttered system and a collaborative, transparent relationship with the regulator to achieve this, as legal defensibility creates roadblocks to full team ownership.

So, what does this mean for directors now?

Reframing due diligence requirements as we evolve to the 'new view' of safety

The director of one company I worked with shared that since doing Safety II and Learning Teams, he could now see how smart and capable his whole team are, where previously he'd seen them as a rough bunch he didn't entirely trust. In another company, after doing Learning Teams, the officers became aware they'd been blind to how much the workers could contribute to solving production problems.

Any transformational journey in an organisation requires understanding where we are heading with the company directors' focus. Change is in the wind for everyone.

In 2023, New Zealand's Business Leaders' Health and Safety Forum published a report on *Better Governance* based on

New View Safety concepts.[2] To read the full report, go to https://www.forum.org.nz/resources/better-governance/. The report draws on legal requirements in Australia and New Zealand, but I'm sure the principles have relevance in other countries. Let's examine insights 1, 2 and 6, to frame where we're headed in this book.

8 KEY INSIGHTS

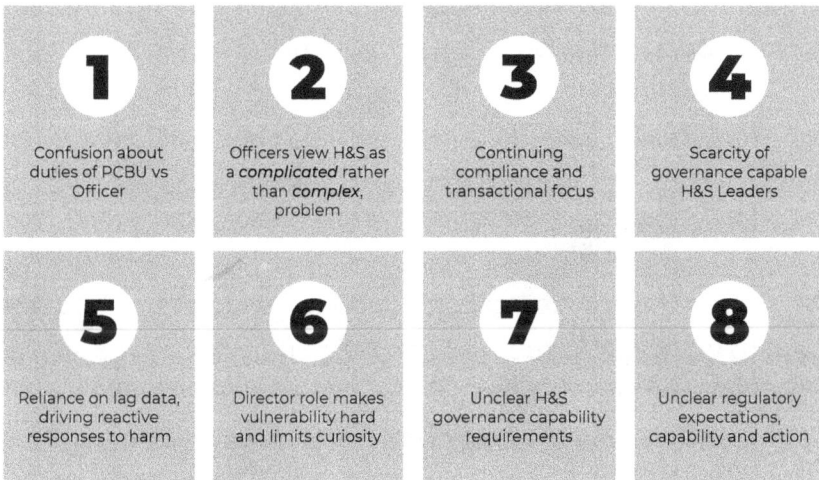

1	**2**	**3**	**4**
Confusion about duties of PCBU vs Officer	Officers view H&S as a *complicated* rather than *complex*, problem	Continuing compliance and transactional focus	Scarcity of governance capable H&S Leaders

5	**6**	**7**	**8**
Reliance on lag data, driving reactive responses to harm	Director role makes vulnerability hard and limits curiosity	Unclear H&S governance capability requirements	Unclear regulatory expectations, capability and action

Figure 3: New Zealand's recent 'Better Governance' findings

Insight One concerns directors' understanding of where their personal liability begins and ends.

In New Zealand, the organisation is liable if someone gets harmed. Individual directors are only legally required to carry out due diligence duties, which is to conduct activities which support the organisation to avoid harm. They must know and understand what's happening in the organisation and resource the teams to keep themselves safe. In addition, all compliance must be met, and monitoring and verification must be undertaken to keep checks and balances in place. We'll explore these more in Chapter Twelve, where we match the duties to the New View of data to collect.

In practice that means transparency over what's going on in the organisation. Directors can't fulfil their duty without this awareness, so part of our job as safety leads is to enable transparency. The only way we can do so is through building trust so people feel safe to tell the truth, despite the power distance created by hierarchy. Usually managers filter what is communicated up and down, partly so they can get the job done without undue interference. This needs to be cut through with new solutions.

What this means is, as safety practitioners, we need to find new ways to work within the hierarchy of our organisations so that the voices are heard, seen and transparent to all. We'll look in Chapter Six at how to do this with directors and leaders of our businesses. So, you become more of a coach and facilitator, and the role of expert giving advice becomes only one part of the picture.

In Company Z, the team shared what they were sensing and the company didn't respond regardless of the good intent. That means the team will be less likely to come forward in the future.

Insight Two builds on this. Safety I, as a term, speaks to viewing safety as a 'complicated' problem, which is where professional expertise has the solutions. If you build a rocket, someone with know-how (an appropriate engineer or technician) is required for the job.

Safety II, however, attempts to deal with the complex nature of our workplaces. In a rocket factory, how the teams work together to get things done is not always predictable (this is a core feature of my next book). Asking how teams adapt successfully to changing day-to-day conditions deals with this complexity. The associated risks are what reporting to directors should be about. These matters must be transparent in the business.

The Company Z teams struggled with resourcing to keep the facility clean. In law, it is clearly the responsibility of the organisation to resource the teams to enable their safety and wellbeing. In New View Safety language, we're talking about the visibility of system vulnerabilities. In this case, appropriate cleaning wasn't being undertaken for reasons beyond workers' control. And for this, we need input from the workers who know the situation.

The Sixth Insight raises the topic of vulnerability. Like all humans, directors are uncomfortable not having all the answers, so they need to develop curiosity. Systems can support this, and leaders will always have the right of veto. There's more on this in Chapter Three.

To fully understand what's happening in the business, directors need frontline workers to have an amplified voice. Yet, as we've seen, these voices have been devalued in our industrial society. In addition, these workers need transparency over decisions made at board level, plus financial and budgetary information to support their ability to 'sense and respond' appropriately.

Starting on this journey can be iterative, from a micro-experiment basis.

In 2023, the winner of NZ's overall industry safety award engaged their work teams in supporting the organisation with capital expenditure prioritisation. Setting aside a sizable budget for improvements and using extensive workshops with the teams, the health and safety representatives were able to decide how the money was spent. The head of safety told me that it took a bit of persuasion to get the project through, but after the first year, the board agreed that it needed to continue.

Dianne Chadwick and her colleagues in the oil and gas industry showed that safe behaviour (and reduced illness and injury) was linked to trust between workers and managers.[3] Some of what must be heard is not always palatable. The truth can be inconvenient and hard to handle, so it requires a courageous culture. Yet trust emerges from this bravery. Increasingly, published academic papers are indicating that high trust in organisations correlates with reduced injuries. This is the way forward. Anecdotally, the front-runner organisations with New View Safety report significant and consistent reductions in worker compensation costs.

A final word: inclusiveness

The workers who are most often dumbed down come from marginalised ethnicities. In New Zealand, this is commonly Pasifika, Māori, Asian and immigrant workers from developing countries.

I worked with a CEO who employed workers from diverse cultures. In his words: 'When an organisation gets bigger, you take some things for granted. To truly understand people, you have to get to know them. You have to get to know their lives and their families, how they live and what ticks people's boxes.'

He went on, 'I think that in some ways we're hell-bent on turning things into a more corporate relationship. So, for me, the work to understand how people are and where they've come from really helps build that relationship. The whole safety side is about relationships.'

Throughout this book, we'll explore case studies where all workers are enabled to contribute to learning and making improvements in organisations. We'll show that humanising people works.

Where to from here

The good news is that every good thing requires patience. Heading into the New View Safety journey takes time. Consistently heading in the desired direction and taking everyone with you can take longer than you might imagine. Think three to five years to achieve what you're visualising now.

Start wherever works for you. There's no one-size-fits-all in this. Learn from others. Team up with like-minded colleagues and form organic innovation groups. There is much thought leadership available to help you develop your approach. Read widely, learn from those who've gone beyond, and join in by creating your own case studies to share with others.

Chapters Two, Three and Four set up the journey with the basic theory, including resilience engineering, how to get your organisation ready to get started, and an understanding of the foundations required for success.

Chapter Five unpacks the change management journey.

Chapter Six is the starting point for engaging your leaders and managers in a new set of human organisational performance principles (HOP) as they support your teams.

Chapters Seven through Eleven explore setting up a Learning Teams programme and framework with cultural competence for all your teams, including contractors.

Chapter Twelve concludes everything we've learnt by starting the conversation on a new way to measure safety and provide worker intelligence to your directors. The aim is to build trust and transparency based on the Learning Teams programme you've implemented.

Finally, once you've developed more trust in your business through these methods, the Epilogue offers a sketch of my second book (coming soon). That book details how to co-design the safety programme with your team and then stretch to supporting teams to self-direct the programme themselves.

Now that you know where we're heading, the first question to address is how safety science underpins our discussions.

CHAPTER 2

Introducing New View Safety

New View Safety improves business performance and has positive spin-offs for all other stakeholders – it's a win-win

Key points

- New science is emerging to guide businesses into a new paradigm.
- Organisational benefits go far beyond safety improvements to building team capacity, improving performance, and power sharing.
- Resilience engineering is the basis for the evolution of our safety programmes.

I once ran a Learning Team with a forestry company that had a frequent issue with the machines that pulled the cut logs out of the forests. The machines often rolled down the hill, costing the owner half a million dollars each year in repairs. They'd developed the machinery to the point where harm to a person was unlikely, but this issue was biting the CEO's pocket and causing lots of downtime.

Most of the time, the teams did a sterling job, navigating a complex job in hilly back country with limited resources and manpower. It's not exactly easy work.

The CEO mandated a process where, rather than blaming the teams for the problems, they were asked how they were handling the job successfully. This may seem illogical under the circumstances, but bear with me.

It meant understanding how the teams handled goal conflicts and navigated difficulties and what they depended on to get the job done. Once the company had transparency over system vulnerabilities and how they were being managed, an upward spiral began.

It turned out that teams were operating within everything the safety management systems told them. That discovery is a clear reminder that humans are not weak points in systems to be controlled. Instead, they are highly adaptable actors who balance unpredictable demands and constraints while keeping the system operating.

The workplace they deal with every day is a dynamic network of interactions and relationships. Individual and collective behaviours mutate and self-organise, corresponding to daily events. The question becomes, 'How are our teams creating success out of our imperfect systems, and how do we support them to create more of it?'

There was far more to it than I can describe here, but in simple terms, it turned out that one forest crew had no problems with machine rollovers. But this wasn't discovered until we got the crews together and started asking good questions.

The crew explained their unique approach to training and developing team members. Handling these machines was quite easy, so newbies often got the task, leaving experienced team members with more technical jobs. This crew created a whole team role around training the newbies. It was this extra tuition and supervision that was making the difference.

Fast forward one year, and this company was happy to report that the rollovers had stopped happening. It saved them half a million dollars a year and gained street cred with their forest client. That's nothing to sneeze at.

As you continue through this book, you'll see a thread through these stories. Who could have thought that better ways of doing safety would also create wins for the business? Looking at this story, it seems obvious. The first step is understanding what all this new lingo is all about.

What is New View Safety?

New View Safety is an umbrella term that encompasses a related group of theories based on new science and academic enquiry. These embrace an understanding of complexity that challenges the Tayloristic thinking of the twentieth century.

If you're in the safety industry, you'll have heard terms such as 'Safety Differently', 'Safety II', 'Human Organisational Performance' (HOP), 'high-reliability organisations' and others. I'll refer more to these schools of thought as I link them to practical implementation.

Despite appearances, these theories have not simply popped up out of nowhere. While the originators are clever people, these 'new ideas' are, in many ways, the synthesis and next logical steps of a raft of new science, new applied theory and other trends that have been bubbling along for decades and longer. They are being utilised in other disciplines beyond safety.

Rather than providing a full rendition of the theory, which you can easily find elsewhere, I'll talk through the practical application and bring the theories in as we go. However, we do need a basic introduction.

I grew up in the residential construction sector, where I became a supervisor at the tender age of twenty-two and learned how operations work in practice. (As far as we knew, I was the first female to do this for New Zealand's tier one construction company in the 1990s.) So when the New View Safety theory first hit the streets, it totally clicked for me. It's about aligning our safety approach to what makes sense in terms of getting the work done.

Perhaps surprisingly, we don't talk much about safety when doing Safety Differently. We mainly talk about work. The idea is that safety naturally emerges when work is done well and smoothly, and is well-organised and well-resourced. There you go, I've just defined New View Safety for you. The rest of this chapter is simply developing the concepts further.

> Essentially, the job of leadership in New View Safety is to develop a better understanding of how work actually happens and give the team tools to work successfully and safely.
>
> Safety is one of the emergent outcomes if the work is going well. Constant reliance on management to make decisions for the team is usually a hindrance.

A colleague recently texted me after running an AI request for a simple, clear description of the new thinking. I laughed at the output, yet marvelled at its cleverness: 'Imagine you're playing catch with a friend. Safety I is like watching out for dropped balls, while Safety II is about figuring out how you throw and catch the ball smoothly every time. Safety II helps us learn from good things to keep everyone safe!' Yep, AI hit the ball out of the park on this one.

Let's return to the story at the start of this chapter. The forestry CEO shared his understanding of the enquiry process: 'We got what the good and bad looked like and merged them. Then we took out the bad, and now we are left with the good.' This new focus dropped their damage rate significantly. Currently, though, most safety systems focus on the bad. Over time, our challenge is to adjust our lens and adapt our systems accordingly.

It's important to remember not to throw the baby out with the bathwater with this new definition. Obviously, we must still provide a safe working environment and work together to follow safe practices. New View Safety is an 'and', not an 'or'. We keep the best of our current approaches and evolve into our potential using a new philosophy that promises an ethical and high-performing organisation.

It's like Russian matryoshka stacking dolls. The critical risk management systems developed during the Safety I era are like the inner doll – still the core ingredient. Safety II approaches build on what we've already developed, adding the next doll while keeping the integrity of Safety I intact.

We still must manage critical risk to ensure safe outcomes. However, we also recognise the importance of what drives team capacity and supports them as they adapt to varying work demands. This way, we understand what resourcing they require to stay safe.

CAPABILITY	Training and know-how
	Team skills
	Physical and psychological capacity
	Manpower
	Understanding of work to be done
	Leadership and culture
	Work planning and organisation
RESOURCES	Time allocated
	Budget allocation
	Supervision/management support
	Consumables
ENVIRONMENT	External factors/stakeholders
	Immediate working environment
	Working conditions
SYSTEMS	Documented systems
	IT systems and functionality
	Work methods and processes
TOOLS AND EQUIPMENT	Appropriateness and availability
	Robustness and ease of use
	Asset design

Table 2: Capacities to consider

It's also important to understand how teams handle demands, such as unplanned variations, work delivery demands, organisational constraints, the impact of goal conflicts, planning and compliance obligations. (Check out page 22 of the Due Diligence Index for more details – we'll discuss the document in Chapter Twelve.)

It may be hard to believe it's all this simple and obvious. Make life easier for the teams? Be more supportive? Collaborate with them? Resource their needs and give them greater ownership? Trust them to make more decisions? Is this what safety is about now? Yep!

Let's illustrate this with an example. A carpentry team told me that constant variations to the job designs by the client, the undermanning of some critical path tasks and issues with who got priority use of cranes to lift gear and materials were causing problems on the job. None of this was covered in their safety register or itemised in the safety audits; however, a combination of these conditions meant they were under constant pressure (an all too familiar story across all our workplaces).

There were other constraints, too. A lack of storage space on an inner city site meant they had to manually handle product and equipment, which increased pressure and reduced wellbeing. The team liked their management, who were kind and respectful, so they didn't want to complain, but they had been sucking up the situation for months.

We sat down to understand this picture and collaborate with the team on the solutions. The team suggested no-brainer operational improvements that didn't cost the company much, made life easier for everyone and reduced the risk of manual handling injuries.

Management were vulnerable in this situation, given that a hierarchical organisation assumes they should have these problems sorted. However, the reality is that management will struggle to be fully aware of issues until a process like this makes them transparent. Time and time again, I've seen relative awareness from managers coupled with a few good surprises. They are usually grateful for the process which spotlights problems they can now solve.

In this case, the site management were great. They listened and instantly implemented fixes based on the team's recommendations.

As my background is in construction, I recognise the challenges the entire industry faces in adopting new ideas. Systems transparency is tough in a complex hierarchical contracting food chain where contractors have to deal with the commercial realities of pleasing their clients in resource-constrained and deadline-driven contexts.

I see directors and leaders of higher tier construction organisations unaware of the system stress absorbed by frontline workers across New Zealand construction sites. The ability for these workers (who are at Todd Conklin's 'sharp end') to have an appropriate 'response' to what they 'sense' is diluted by layers of site and project management, engineers and consultants, and leaders and directors of the various layers of management, and directors of the companies involved.

The situation is often exacerbated by the diversity of cultures within the workforce, who can struggle to challenge authority when needed. The challenge for the industry is finding real solutions when directors and leaders sometimes see frontline workers as low-skilled and less capable of contributing to decision-making. That's a fallacy the New View debunks.

The project-by-project nature of construction work means it has higher variability and demands than many other types of work. To build resilience, we need deep understanding of what drives team capacity. Questions should also be asked about what would dampen variability.

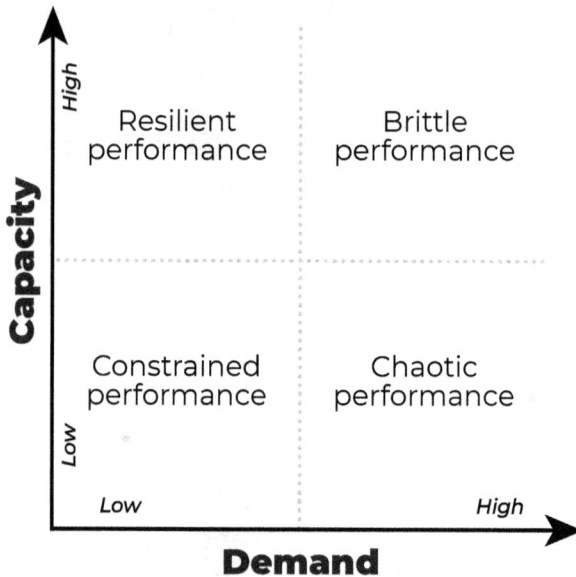

Figure 4: A New View of work demands – diagram courtesy of Art of Work

In a nutshell, the New View Safety theories provide the roadmap to rethinking how we view and arrange better safety at work. There's new jargon to learn, but, in practice, none of it is more complicated than the story I've just told. We just need to learn how to systematise this into our businesses.

In this instance, it took learning sessions to bring the truth to light. The structured process supported the carpentry team's need to 'sense and respond'. It let their insights be heard and provided a situation where management could respond well to matters that needed to be sorted. The opportunity arose for the client who provided project management and monitoring to join the learning and deepen their understanding.

Every safety professional, operations manager and senior leader needs a full working knowledge of the latest developments in safety science and current methods for implementing these simple principles.

The unspoken challenge tends to be that imperfections in the systems are unclear. This is due to the politicisation of how we work together and is driven by the hierarchical way we organise ourselves. This must be addressed effectively, and it's not easy. Hence, the opportunities that lie in following the roadmap and wisdom shared in this book and the next.

The game changer: changing the definition of safety

There are clear synergies between the application of recent neuroscience discoveries and the New View Safety movement. That matters because it frames the change in thinking we need to adopt.

Among many revelations, we've learned that what we focus on is embedded through a process called neuroplasticity. Our thoughts, behaviours and beliefs become autopilot and create our behaviour over time. Given that what we focus on amplifies, if we shift our focus, we can enable entirely new patterns of behaviour.

In his body of work (starting with the paper *The Tale of Two Safeties*[4]), Erik Hollnagel shows that the focus of our safety programme is the source of our problems. He says we are constantly focusing on avoiding anything negative. We look for what could go wrong. We count the times when things go wrong or nearly go wrong. We are so conditioned to this lens that we don't even question it.

Motorsport offers a great example of the importance of this point. If you make a mistake and lose control, the worst thing you can do is look at the wall, barrier or car you're about to hit. In that moment, you must switch your focus on a 'gap' to get into (the best direction to travel in). Somehow, your brain will find a way to get into that gap, if it is possible, within your current technology and skill set. All good (and surviving) race car drivers know this.

In organisations, our focus needs to shift to building strong understanding and systems for our teams to get into the gap. In the parlance of New View Safety, we improve the teams' capacity to adapt to whatever comes along in the work day, meaning they become more resilient. There, it's that simple. As they say, it's not rocket science. And this is what I've been leading up to all along. The rest is easy.

New View Safety is still in its infancy. Just like developing new neural pathways, we need to reshape our entire systems and processes so they are built on this new language and philosophy. That is the journey we're all on.

When you ask residential construction site teams what drives team resilience, they will talk about the various conditions and constraints and what supports success on the job. Typically, they include:

Teamwork. The ability to build a cohesive team and good communication, including through daily prestarts.

Job planning and project management. Deadline pressures and goal conflicts that inclement weather can exacerbate.

Varying levels of skills, experience and competency. Senior team members are often stretched thin due to skill shortages.

Resourcing and supervision support. The job goes well with sufficient resource and support.

Nature of working conditions. The hard physical work in construction requires mental stamina and focus. Coupled with teamwork and job planning, this can help or hinder performance.

So, what should our safety programmes focus on? We know that we spend a lot of time telling people what not to do. We ask them to tell us anything that is going wrong. Instead, we should focus on initiatives that improve conditions and support further resilience of the teams. Usually, this is seen as the domain of operations rather than safety. However, we just can't divorce safety from how we work, which is what we've been doing for too long.

It's rather a radical new way of looking at safety, isn't it? We're so used to thinking of safety as a bunch of rules. You won't find the five points listed above on audit sheets, except perhaps the fifth when the safety profession looks at human factors.

Interestingly, in almost every Learning Team I've ever done, workers tell me their teamwork is the number one contributor to success and safety. When they work together well, the job goes well. That makes total sense when you think about it. Therefore, every safety team should be focused on supporting teamwork as their priority. Yet I rarely hear it mentioned; instead, we focus on rules, procedures and paperwork.

As with motorsport, this is the gap we need to get into. We need our organisations to pay more attention to and develop initiatives to support what the teams tell us provide safety for them. Our leaders should ask questions around these conditions, and our directors should resource improvements. For this, we need transparency, which in turn requires trusting the teams to tell us the truth, even in circumstances where it's difficult to do so.

We don't want to lose the goodwill or good work. Everything that could go wrong and the barriers to preventing it are still part of the story – they're just not the whole story. We need to ask how things can go well and how to get more of that. Erik Hollnagel and other safety scientists teach that this should be our primary focus.

By way of a definition, Karl Weick describes safety as a 'dynamic event', i.e., something that must be created constantly and continuously. The basis for this creation is that which makes up everyday work.

Safety professionals have been taught to 'reduce' safety to what we can write on a piece of paper, a definable control we can tangibly identify, or a rule or edict we can deliver to our teams. The scientific term for this is reductionism.

Of course, these things still support safety, but they are why Safety I (the traditional view) gets a bad rap. It fails to appreciate the holistic nature of how safety occurs in practice. Safety can absolutely happen without paperwork, controls or rules. In this sense, it is dynamic, occurring in every moment, often for reasons beyond our control as managers and usually via the teamwork workers describe.

So, your new job as a safety lead or operations manager is to enable team resilience rather than obsessively controlling their actions, which is the legacy of behaviour-based safety philosophy. A lot goes into the how – enough to fill this book and the next. However, what's most important is focusing on what drives success.

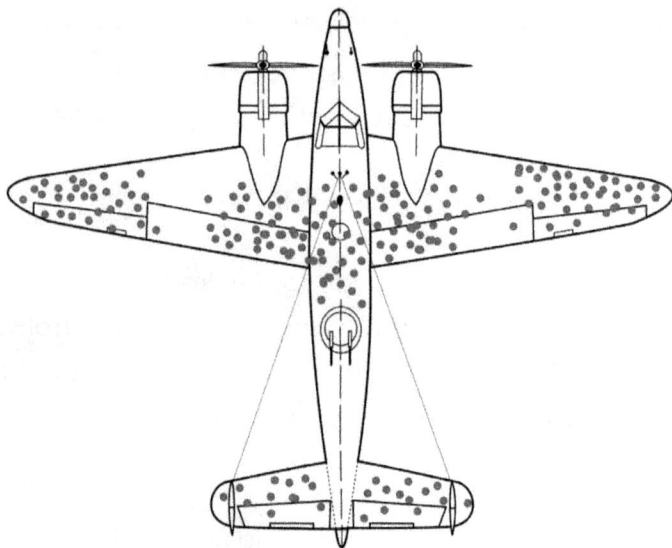

Figure 5: World War II fighter plane

There's a story that is often told to explain this. During World War II, the Allied forces noticed that returning aircraft that had been heavily damaged by enemy fire often had bullet holes in the same areas. At first, they assumed these areas were the most vulnerable and reinforced them to better protect the aircraft and crew. Much as our current safety systems operate.

However, some genius thought to look at areas on the planes with the least damage and asked further questions about how they'd made it home successfully. He hypothesised that reinforcing these areas might mean all the planes could withstand the bullets. Sure enough, it worked.

Similarly, Safety II seeks to understand how teams are currently creating success despite the imperfect systems we give them. We aim to strengthen those elements instead of only examining failure.

As with the WWII planes that made it home safely, organisations do not easily see what is creating success. There's usually no need to if everything is going well and there are other matters to attend to. So, we must look hard and train organisations to focus on these things when our current systems don't. That takes time and effort.

It takes time and repetition to establish a new personal habit. Coaching organisations into new thinking will require the same effort.

In residential construction, work teams often tell me that issues could be nipped in the bud if they had more bottom-up input into project management. They could provide direct feedback on what's working and what's not. They say management often pull team members off to other jobs, interrupting the team dynamic. Two-way communication would make the workflow a lot better.

We reduce safety improvement to a tirade of corrective actions that may or may not reflect the complex interrelationships of what supports and enables the teams. Focusing on what creates success will better support teams in complex working environments.

Current safety management systems are woefully inadequate to achieve this. The safety function is not set up to support it either. That includes how we measure safety and the questions asked by senior management and governance, who are still mainly focused on control. Director training currently teaches them that their job is to maintain control. We must inform our directors of the shift in safety science and the reframing to supporting capacity and resilience.

Avoid the drift to failure

Before introducing New View Safety to a new company, I like to do a short site visit to meet some of the people. On one such visit, a construction worker kindly welcomed me as I walked around the site. When I explained the reason for my visit, he said, 'Well, the thing is, we have to adapt to the conditions we face on the day'.

He unknowingly quoted Erik Hollnagel, Sidney Dekker and all the other highly respected internationally renowned academics in their exact language. As I said, this is not rocket science. When our systems listen to the teams, we can ask how they need us to support them to make better decisions in varying conditions. The systems adaptation required means understanding these conditions and driving continuous improvement.

Workers learn to overcome design flaws and poor planning because they recognise the actual demands and adjust their performance accordingly. They interpret and apply procedures to match the conditions — just as that perceptive worker had pointed out.

> The teams detect and correct when something goes wrong or is about to go wrong. They intervene before the situation worsens. This results in performance variability. Not negatively as a deviation from a norm or standard, but positively, in that variability represents adjustments that create safety and productivity. We call this 'positive adaption'.

Workers know they are doing this all day, every day. And it's why they get so fed up with safety programmes which neither acknowledge nor respect this reality. We interpret it as a negative attitude towards safety when, in fact, we are doing a poor job of listening.

We need to learn how teams handle the competing demands of keeping safe, ensuring the work supports economic objectives and tackling the necessary work delivery demands. Conditions and constraints cause variability, including (but not limited to) those in the following list:

- production pressure
- unclear signals
- local factors
- adaption
- errors
- resource constraints
- change in plans
- flawed processes

- design shortcomings
- personal factors
- poor communication
- goal conflicts and tradeoffs
- past success
- incomplete procedures.

The challenge becomes how organisations introduce a robust enough programme to ensure they pick up 'drift' from good practice before it occurs.[5] That means learning how to listen to weak signals (the subtle signs that something's not right) and dynamically and continuously understanding what's going on in organisations.

To be clear, we're talking about listening to our teams more than telling them what to do – a departure from current safety programmes. It takes trust for teams to tell us what's going on. Often, we see a legacy of safety programmes that rely on discipline, a history of union battles, parent-child ways of communicating and systems that see people as the problem.

Shifting our understanding of expertise

A better understanding of what drives success (despite variability) means embracing the idea that workers are the experts in their work, and the risks. Of course, they sometimes need subject matter expertise as support, but we must share power with those who have the best wisdom and knowledge.

Before you reject this idea, check the definition of expert. It's a person who is wise through experience. If you're doing something well, often enough, you'll become the wise one. That means the frontline is usually best placed to solve the problem.

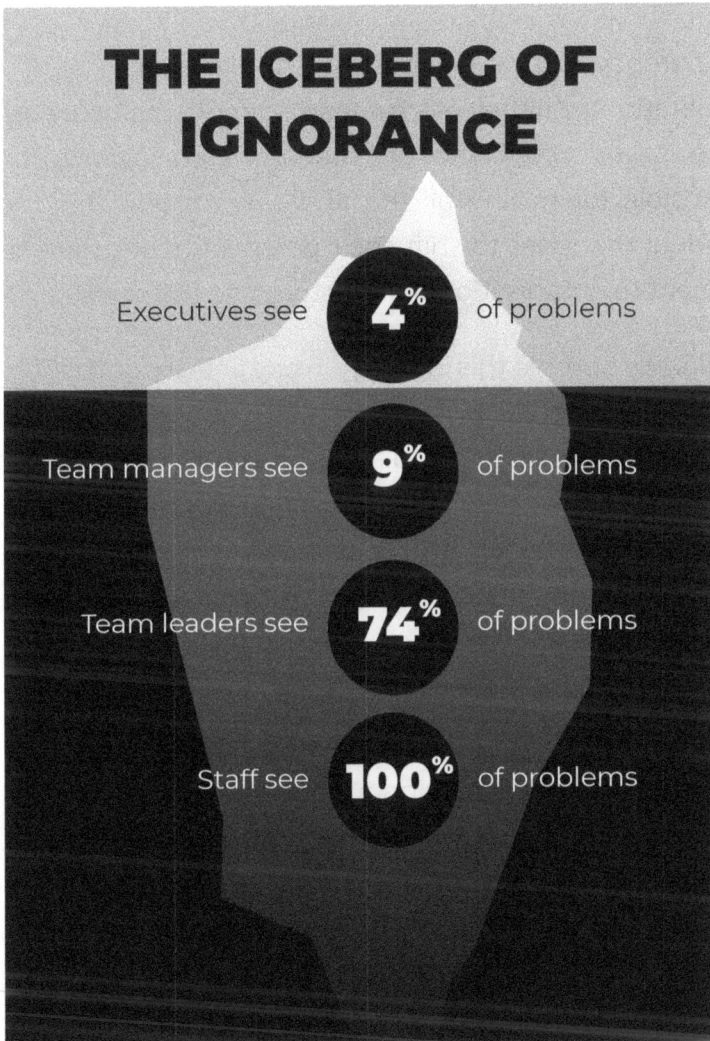

THE ICEBERG OF IGNORANCE

Executives see **4**% of problems

Team managers see **9**% of problems

Team leaders see **74**% of problems

Staff see **100**% of problems

Figure 6: Recognising the expertise of the work teams

Don't read too much into the statistics in Figure 6; just capture the reality of its intention. The New View Safety community has co-opted a cumbersome yet descriptive label: Work as Done versus Work as Imagined. The idea is that those at the top of the iceberg tend to imagine the reality of day-to-day work because they're not practically immersed in it every hour of the day.

Transparency allows a manager to develop solutions in collaboration with their team, meaning more brains are applied to the issues. Often, though, managers are used to feeling responsible for having all the answers. It's one thing for an individual manager to shift their perspective; it's even harder for an entire organisation or industry to do the same.

You already operate this ideology in any activity that recognises your teams' expertise. For you, implementing New View Safety is about doing more of it and redesigning your systems to bake in this way of operating. When you change your systems, these principles become an embedded way of working together rather than relying on the strengths or capabilities of individuals who naturally work this way.

There's a lot to learn.

The first task is to embark on a leadership journey to examine the all-important 'how' of harnessing expertise within your organisation. We're after a mindset of curiosity and better listening to build the trust necessary for work teams to share their insights honestly and without fear of reprisal. For many leaders who are used to being rewarded for good answers to problems, that means being ready to make change.

CHAPTER 3

The Readiness Test

Change involves adopting progressive ideals when resistance to change prevails

Key points

- Businesses are at varying stages of readiness to head into New View Safety territory.

- Assess the readiness of your leadership and organisation to determine your strategy.

- Test readiness and deal with varying levels with different individuals. The answers lie in the paradox between pacing change and overcoming the fear of losing control.

There's an important hurdle to clear before we get into the 'how'. Readiness.

Given that New View Safety is known to improve performance and create win-win solutions, why is there still resistance? After all, it all makes good logical sense and benefits everyone. If you've been following the New View Safety movement, you know it's not always easy to get new ideas implemented. The issue is readiness. Why, then, are some not ready at first?

New View Safety involves a shift from the Western emphasis on individuality to a collective culture. We are dealing with the legacy of management fads, poor industrial relations and accumulated distrust. Moving on from this requires commitment. It takes empathy and understanding to start listening to teams.

Management has long been rewarded for traditional, hard-driving, even forceful behaviours, where identifying and having the answers to problems is valued. It's part of the entrenched social fabric of how we work together.

Furthermore, a little vulnerability is required from managers and leaders when the reality of how workers are adapting to imperfect systems becomes visible. That doesn't come easy in the context of modern reward systems and management philosophies.

Leaders must tackle vulnerability to build the necessary transparency to enable shared responsibility (the holy grail of change). We can't increase team resilience without building trust; transparency is the means to get there.

Being vulnerable is not easy, especially for those responsible for leading organisations where failure can have serious consequences. It's human nature to want your identity intact, so hold any reluctance without judgement and create an environment where there is readiness to create change.

I've written this chapter from the perspective of someone getting started with New View Safety; however, the principles apply whether you are expanding, renewing or invigorating your Safety II implementation.

What drives readiness?

I've seen a variety of reasons why an organisation is ready, and it very much depends on context. Here are some common examples.

- A serious incident or series of incidents has happened, and there is an honest reflection that change is needed.
- There is a problem area in part of the organisation and the New View approach is seen as having the potential to support performance improvements.
- There's a sincere and visionary desire to be progressive and ahead of the curve.
- The need is client-driven or imposed by a stakeholder.

I have also seen companies that include New View Safety in their strategy and say they want to adapt to new ways of working together because they've heard about it and want to jump on the bandwagon. Often, though, they don't fully understand what it takes or what they're getting into, so New View tools and methods are not used as effectively as they could straight away. Genuine success requires being up for the challenge of changing how we work together.

One company I worked with, in a high-risk primary industry, declared they wanted to use the New View methodology but were difficult to deal with when it came to getting started. Initially, they responded well; however, a deeper look showed that the leadership was not ready to listen to their teams.

There was a historical element, including conflict with unions. In this instance, they were giving mixed signals of readiness. The best thing was to provide some small experiments to wrestle with and get them going slowly and steadily. We'll cover how to do this in Chapter Five.

Is it resistance or a lack of readiness?

A colleague in my industry regularly sought my advice, lamenting that her organisation was not ready to trial New View Safety. To put it bluntly (and I see this often), simply asking your leadership if they'd like to try new ideas is not a readiness test.

Assessing and responding to readiness is as important as the New View Safety activities themselves. Make no mistake, if there is no readiness, you are better off waiting — as painful as that may be. It's not individual readiness; it's the readiness of the ecosystem that is important.

You must craft a readiness test and determine if your ecosystem (leadership, teams, industry and other stakeholders) will support new ideas. While a fantastic coffee companion, my friend interpreted resistance as a lack of readiness. These are two very different things. Testing readiness effectively and getting the nuance of understanding risk appetite sets you up for success.

> The path to implementing New View Safety is currently littered with partial or limited success. Getting your head around assessing readiness and pitching your change programme accordingly saves time, angst, wastage of resources and setbacks.

The answer to the readiness test can be that more patience or a savvier approach is required. More preparation around understanding New View Safety is needed, or most commonly, working out where the real resistance lies and having the courage to face and deal with unpleasant visible opposition. One clever client, who was struggling to get the head of operations on board, managed to appeal to the directors for support. They saw the benefits and were happy to throw their weight behind a new approach.

I've seen many teams start New View approaches and spend significant money on initiatives yet not get the success they wanted. As a result, they give up or get distracted with other work. Often, it's because they didn't prepare by understanding readiness and working accordingly.

The good news is that my friend did take up the challenge and had an honest conversation with her CEO. She said, 'I wish to trial these ideas, and I'm willing to wear the consequences if they fail'. He respected her enthusiasm and gave her the go-ahead to do a small trial and assess the boundaries of readiness.

Partial readiness

Your organisation may simultaneously be ready to trial some New View Safety ideas and not ready for other aspects of the new approaches. That will likely emerge during the initial discussions and trials. Given this mixed appetite, pay attention and adapt your rollout approach.

This situation is best illustrated with an example. A civil construction company wanted to try New View Safety to take a fresh look at an entrenched industry issue: repeat service strikes (where digging operations hit electrical, gas or other services inadvertently). They had been doing thorough and competent investigations, had identified human error as the main problem, and issued written warnings where fixed rules had been breached. Yet things weren't improving.

It's an easy trap to fall into. Using discipline to control behaviour and achieve conformity to well-considered rules is an attractive approach. This company wanted to find new information while retaining the integrity of the current system.

To this end, they ran a Human Factors session with leadership and a Learning Team with the work crews. The Human Factors session reinforced Safety I language, such as 'error' and 'violation' as the cause of issues, although they introduced a fairer lens on the discipline used. The Learning Team, however, sought to understand the whole situation from the workers' perspective. The blend of methods showed mixed readiness. Over time, they succeeded with Learning Teams, their readiness grew and they took on more courageous New View Safety projects.

Trialling readiness

When determining readiness, it's important to test or trial how ready an organisation is for change. As a safety lead, you don't want to overshoot the mark or underestimate the stretch you'll be capable of.

Table ideas and see how they are received to gauge what's possible. In Chapter Five, we'll talk about running experiments and trials that test readiness and give guidance for the appropriate cadence of change.

An organisation came to me wanting to get started on New View Safety, yet struggling to get traction despite repeated attempts. It was clear that they were failing the readiness test. I needed to work out whether they were dealing with resistance effectively. I also needed to set them up for success when they were ready – if it wasn't imminent.

We dealt with the initial resistance by determining who would effectively champion a readiness test. It needed to be someone who could see potential in the concepts and had the authority to make it happen.

In this instance, it was the head of operations, so we designed a small, cheap experiment aimed at the board of directors. We needed to show that listening to their teams would provide insights that would clearly improve operational efficiency.

I engaged the team in the concept of this readiness test, and they duly provided evidence of aspects of production not going that well. If they still weren't on board, there was a readiness issue. Sure enough, the board said they were interested in New View Safety, but the timing was wrong. It was a case of 'not now', so we devised strategies to lead them to future readiness.

Holding the space – building readiness

Sometimes, your leadership team aren't ready. Many organisations are still relatively conservative. That means a strong focus on exacting safety management systems prevails, and leadership prefers a traditional top-down approach.

Let's face it; the New View Safety journey involves a mindset shift that isn't always an easy sell to those leading your organisation. It requires letting go of control, trusting the team, and sharing power.

In this case, it's best to assume 'No' means 'Not now' and prepare the way in advance. The following strategies work in this case.

Priming

Marketers and addiction specialists know this strategy well. The best way to get behaviour change or mindset shift is intentional exposure to new ideas over time. Momentum builds to a tipping point like a dripping tap forming a puddle. It can take the form of giving your leadership pre-reading, podcasts, videos or case studies. Tell them about the many blue-chip companies that are now starting this journey.

This process is all about taking the necessary time and having a game plan. The preparation and priming process is as important as the change effort itself. One client who experienced great success with New View Safety was very effective at using this technique.

He figured what in the New View Safety literature and resources would work with his general manager and fed him a steady diet of material. The GM devoured the material and became the leader of New View Safety in the organisation. Usually, the safety team leads the process, but in this case, the safety lead adopted the position of coach and facilitator – as it should be.

What's in it for me?

New View Safety is based on a win-win ethos for building better work and relationships in businesses. I have two main approaches to getting started. The first is finding a low-hanging opportunity as a starting point. Ideally, it's a cost saver that's a no-brainer with little legal risk. The forestry rollover example in Chapter Two was a case in point.

My second approach is to figure out what resonates with the CEO's intellectual framework. CEOs appreciate pathways to new vistas based on their personal curiosity and growth plans. Often, they find the concepts and theory of New View Safety fascinating and are keen to learn more from personal interest.

One CEO I connected with didn't immediately want to rock the boat with New View Safety. It was mid-COVID, and it just wasn't a priority for him. Then I mentioned Appreciative Enquiry, a subject he was fascinated with, and his response became, 'Why didn't you talk about this sooner?'.

Another CEO was completing a research paper on the application of systems thinking. He was always keen to learn more, and if I kept up his appetite for new theories, he was always prepared to back another New View Safety project.

Building credibility

A vital aspect of building acceptance is establishing the credibility of New View ideas. It can all sound somewhat progressive and fruity to some leaders, but it's grounded in the latest science. Find organisations and institutions that your leaders respect. Identify those that are introducing new approaches and leverage their positioning.

Good examples in New Zealand include Worksafe, NZ Business Leaders Forum, NZISM, Safeguard, CHASANZ and countless companies that are trialling the ideas. Many of these organisations bring in guest speakers or ask leaders to attend workshops and other events.

Having an experience

Another technique is working with the leadership team on a personal issue. Experiencing a Safety Differently process often gives insights that sharing theory can't elicit. An example is conducting a Learning Team on what baseline capability is needed for New View Safety to become effective as a new approach to operational learning.

Boards and executives fail to understand the reality of work

Another technique to building readiness is demonstrating that unless your organisation uses new methods with their safety approach, they are potentially exposing themselves legally. I avoid this approach because using fear as a motivator generally limits the extent of your success. However, it's an option if you're struggling.

Due diligence requirements expect officers to know and understand safety in the business. They can set themselves up for legal risk if they are unaware of how work is being conducted. Traditional methods often miss how teams handle the messy reality of work, giving directors a false sense of security. The transparency we're talking about helps them meet their legal requirements.

Another organisation in a high-risk industry wanted to trial Learning Teams. During the process, they developed enough trust to learn that the permit-to-work process they had built to satisfy a mandated requirement for a safety case was not being used as intended. Several layers of paperwork were required, which had become too onerous to be practical, so the team created a workaround. They would answer the questions on the form in ways that would override the triggers to fill out the next layer of paperwork.

This layering effect occurred after several Worksafe prosecutions caused the knee-jerk reaction of adding more and more administrative controls. Sound familiar? Leadership was blissfully unaware of these workarounds. Needless to say, the processes were soon improved.

Having passed the readiness test, you're itching to make headway into adapting to the New View Safety ideas. First, though, let's look at the foundations of success. In the next couple of chapters, we will figure out what groundwork needs to be laid.

CHAPTER 4

Foundations of Success

Ensuring minimum safety management standards are met to support evolution

Key points

- The importance of managing critical risk – aka the 'shit that can kill you' – is still paramount. The purpose of your safety approach is to enable teams to fail safely.
- Doing a great job of critical risk management is the central purpose of safety management regardless of traditional or new approaches.

'Most companies in New Zealand don't have the basics in place, and that's why I don't advise them to use the Safety Differently approaches.' That was the reply from a safety consultant colleague a couple of years ago following a discussion on the new philosophies.

It is typical of the and/or thinking I still hear from those questioning the New View promises.

I want to say this: whatever we do around safety innovation, whether insights from a micro-learning exercise or conversation in the wider debate, we should listen to the naysayers. They have gems to share that are important to the new ideas.

The point is, my colleague is right. We must have basic standards in place. The problem is that I have yet to see a company with its house fully in order, so being good enough first is an elusive goal. It's driven by either fear or a misunderstanding of what's involved in having a go at the New View.

Besides, the New View way often solves the very problem you are bashing your head over with traditional approaches.

We need to get away from black-and-white and/or thinking. We are aiming to *evolve and adapt* from our current approaches. Remember the Russian dolls? At the start, we add new approaches, or layers, to the current mix.

The reality is twofold.

New View Safety is still about good safety management; it's simply a different method for achieving the same aim. Improvement via either method is desirable.

That said, there is no excuse for poor safety management; putting your team at unnecessary risk is unacceptable in any situation. Using whatever method is chosen to make improvements is always required – even if it is totally old school.

The second problem with my colleague's thinking is the influence of beliefs. Neuroscience has conclusively shown that our beliefs shape our thinking, which, in turn, influences our behaviour. Alongside good safety and resourcing, we must consider what beliefs will set us up for success based on what the new science teaches us.

My colleague wouldn't have told me this, but he didn't believe that work teams across the country could step up to what New View Safety says they are capable of. When the older systems have been pointing at what people do wrong for so long, it's no wonder many still see things from this perspective.

Safety leadership and maintaining best practice

Much has been written, taught, and expounded about safety leadership and critical risk management. My intention is merely to say that this is a necessary condition.

The effectiveness of Safety I has been built around the need for good safety leadership, which is, in simple terms, caring about people and doing the honourable thing about making the right decision to prioritise safety in any necessary situation. Most organisations nowadays understand these principles clearly.

It has been the stuff of the last couple of decades, and many noble advocates have done the hard yards to shift from an era where caring had to be legislated because of those who didn't. This era has essentially permitted New View Safety to become a concept worth following.

If in doubt, ask leaders in what areas of their organisation they would be happy for their (hypothetical or real) children to work. If they rule out particular departments or processes, that's a telltale sign that the organisation needs to care more about the teams in that area.

Obviously, meeting basic safety requirements must be the highest priority. Our regulatory bodies and many other institutions, including businesses, have worked hard over the decades to improve workplace standards. Your organisation must continually strive to meet these. And, of course, the standards are constantly evolving.

A good example is when a provider of engineered stone for kitchens and other interior fittings took the risks of silica dust very seriously and proactively partnered with industry stakeholders to develop new solutions. They took the stance of improving their risk controls while informing and educating the rest of the industry. This is good ethics in practice.

A cornerstone of your safety programme is collaboration and partnerships with stakeholders, including regulators, industry bodies and safety forums supporting New View Safety development. Contributing to the wider development of safety innovation and sharing case studies also creates value for your organisation.

Critical risk management – supporting your teams to fail safely

I have always loved the concept of supporting teams to 'fail safely'. It reframes safety simply and effectively. Mistakes and failures are a natural part of how we learn and adapt as humans, so our systems need to cushion any potential blow. I regularly explain this to teams by playing a video of a high-speed American NASCAR car crash.

In one great video, the field is driving in a bunch at around 300km/hour when a car clips another and becomes airborne. The superior engineering of the cars means everyone walks away unharmed – at 300km/hour. A bunch of paperwork didn't pull that off. That video literally drives the point home.

That point was clear in Safety I and is even more apparent in the New View. You must pay attention to your critical risks, aka Todd Conklin's 'shit that can kill you' (STKY). And you must resource continuous improvement in managing these. Todd gave a great example during his 2019 New Zealand roadshow tour, saying, 'It's better to break your arm than die'.

We get so obsessed with the minor things that we often take our eyes off the shit that really matters. Shifting from an obsession with preventing failure to an obsession with failing safely (a success lens) makes a lot of sense. Presenting to your team about these concepts is a great way to introduce New View Safety to your organisation because it resonates.

The rest of the story is that you need to focus on ensuring the controls for your critical risks are the best they can be – which is an aspect of building team capacity. Like the motorsport example, they should be as high up the hierarchy of control as possible. Paperwork, rules and procedures should be needed less. Use new technology, engineer out risk and support your team to perform comfortably and effectively on the job.

Todd Conklin tells the story of a manufacturing facility that had trouble with people losing fingers because of hazardous pinch points in a particular process. The endless admonitions to keep fingers out of the way hadn't worked. So they got innovative and designed new gloves that became fingerless when caught in the pinch points rather than causing damage to a real hand. It took some masterful sewing to achieve this simple yet effective way to create the capacity to fail.

> We also need to look at the adaptive capacity of the controls. Do they align with the work? Are they adaptable when work varies to support resilience?

There's more on this in the Freedom within a Framework section in Chapter Eleven, where we'll define critical controls and more. I will cover this extensively as we explore co-designed and team-managed approaches in 'Evolving the New View of Safety'. At these stages of development of Safety Differently, user-centred critical control approaches are better at supporting team resilience. Giving authority over critical controls in a high trust and ownership environment with a robust performance testing and verification programme means ensuring teams can adapt to variability more safely than before, enabling teams to fail safely if required.

For now, we must ensure we have adequate monitoring and verification processes. Favour useful Safety I style approaches instead of over-the-top reliance on paperwork, excessive rules and discipline to keep people safe.

Line lead safety

Whenever I stepped into a new safety lead role, the first thing I did was have a good chat with the head of operations. I explained assertively, calmly and without budging that safety was their responsibility. I would support, guide and lead where necessary, but it wasn't my job.

The person who leads operations cannot silo safety away from the work. The two go hand-in-hand and cannot be separated. Leadership and operational leadership are the direct influencers and are ultimately responsible for safety. Their day-to-day decisions have more impact on safety than any functional team ever could.

A common challenge is where the organisation has a people and culture team and the safety lead reports to the head of HR. Remember that HR can't directly influence the work either.

I still see many senior safety professionals responsible for leading programmes in safety-critical industries, reporting to a people and culture lead. While the HR people may be well-intentioned and supportive of progressive programmes, they exacerbate the silo effect. In this situation, safety often becomes initiative-based rather than deepening the alignment with work.

Ideally, the safety reporting line should be to the CEO or general manager of the organisation. If this hasn't been established, hold a weekly or monthly one-on-one to ensure they hear safety intelligence from the work teams as directly as possible.

In a safety lead role, I asked permission from my boss (the head of operations) to have a monthly one-on-one with his boss, the CEO. My boss enabled transparency of safety-critical information direct to the guy who held all the purse strings. He was willing to be vulnerable and allow the necessary flow of intel to go where it should. I also supported the process by coaching work teams on how to present capex business cases directly to the CEO and head of ops.

The importance of beliefs

Neuroscience tells us that great actions and outcomes come from the right beliefs. As you saw in Chapter One, new concepts require a different view of how we create safety in our workplaces. The starting point is beliefs.

One organisation I know well has had difficulty embracing the change heralded by academic speakers and early case studies. Resourcing, commitment to good practice, and critical risk focus have been hallmarks of their safety programme. However, a lingering insistence on believing that 'all injuries are preventable' has hindered their ability to see new ways of engaging their teams.

This assumption leans on behaviour-based safety theory, which rewards good behaviour and punishes poor behaviour. The organisation's programme stipulates that any deviation from agreed rules and procedures would result in discipline. This view is based on the belief that the person is not showing a caring attitude toward the company and their teammates.

While these approaches may come from noble sentiments (and caring for each other is certainly an imperative aim in corporate culture), these beliefs created a fixed mindset and an unwillingness to experiment. This period in the company's safety journey has been very frustrating for many operational and safety staff who could see that new assumptions were required.

Let's turn to the beliefs that set a foundation for success so we can move quickly into results with New View approaches.

The four principles of Safety Differently

One of the key tasks in creating a new way of thinking about safety is to be willing to view safety and our work teams from a different perspective. The following assumptions will support success as you start your New View Safety journey.

- Safety is not defined by the absence of accidents but by the presence of capacity.

- Workers aren't the problem; workers are the problem-solvers.

- We don't constrain workers to create safety, we ask workers what they need to do to work safely, reliably and productively.

- Safety doesn't prevent bad things from happening; safety ensures good things happen while workers work in complex and adaptive work environments.

The safety approaches of the last twenty to thirty years have drilled into the leaders the importance of taking safety seriously. They have been taught to plug the holes, rigorously look for what could go wrong and act to avoid that. It has become an obsession in businesses.

Adopting these new principles involves a paradigm shift. Leaders need to start focusing on different things. How can we support teams to work reliably, successfully and safely under variable conditions? What will create the capacity to be resilient under those conditions? We want to know what support our teams will need to achieve that.

It requires an underlying belief that teams are capable of making good decisions and that it's our job to support them. We'll discuss more of this in Chapter Six where we look at setting up leadership with principles they need to support your journey.

Being open to the teams' perspective

Your leadership must be prepared to truly listen to the teams to support these new beliefs. The starting point is awareness of the current state of play. Constructing an eye-opening survey is a great way to start your New View journey. Running a short 'blame to learn' survey can be quick and dirty. Choose a cross-section of people across your frontline teams and ask the following question.

Where is the company on the blame/learn continuum (Figure 7) when something goes wrong?

BLAME / LEARN CONTINUUM

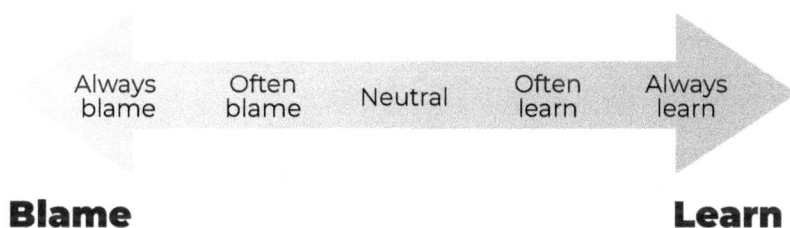

Always blame	Often blame	Neutral	Often learn	Always learn

Blame   **Learn**

Figure 7: The blame/learn continuum

Every company I've run this initial survey with tends to experience a moment of truth. When workers and leadership are asked the question, each group's results are always different. Leadership has to face the reality that the work teams do not always feel their good intentions.

In every instance, workers feel they are unfairly blamed, and that leaders are oblivious to the gap between what they try to do and how it is perceived. It's not about personalities; it's about current systems and beliefs, which tend to lead to finger-pointing.

Another organisation ran a simple exercise to show what the work teams were dealing with. They asked the teams what safety stuff they'd like to *stop*, what they'd like to *start* and what was worth *continuing*. An exercise like this provides qualitative information that can quickly highlight issues.

This willingness to listen and engage will set you up for success. Both exercises prepare you for the change journey, which we'll explore next.

CHAPTER 5

Taking the Change Management Journey

The New View journey involves iterative change

Key points

- Adapting to New View Safety is a hero's journey – change is never easy.
- Keep building and growing your New View Safety journey step-by-step.
- Use safe-to-fail experiments to trial contemporary safety methods.

At the time of writing this book, one of my clients has recently won New Zealand's Safety II national award. An awesome team, they say the secret to their success is starting small and being prepared to make mistakes. They say it's all about supportive leadership. Their leaders faced initial resistance and feared letting go of control but were ready for change. They realised the status quo was unacceptable and could see the new wave coming. They wanted to be ahead of that wave.

So, how do we handle the journey of change? Adapting to New View Safety can be a real challenge, but, like a snowball, it gathers momentum.

Overcoming initial resistance is not for the faint-hearted. There are plenty of tactics, but the most important is showing leadership in the face of fear. I have the privilege of mentoring and coaching people through this process as a full-time gig; however, I've been through it myself and know what it feels like.

In my first Safety II project, I successfully trialled the ideas. The second project was a full-scale pilot of Safety II with an innovation team charged with delivering a multi-million dollar proof-of-concept for an engineering housing system. We used the principles as far as we could.

In a perfect trifecta, I then joined an organisation that was shifting from successful start-up to full-scale production. I used Safety II to fully design the safety framework, which is still embedded today. These three projects occurred between 2015 and 2019 when Safety II concepts were still new and extensively rebuffed as likely a passing fad.

In each situation, I led from the front with ideas (proposed by academics) that weren't yet fully tested in live settings. While I worked in environments with catastrophic risks, I believed the new methods could work based on my previous successes each time. More importantly, I've always believed in the capability and potential of the teams I've worked with. Harnessing their potential was the only thing on my mind. I knew giving the teams power would enable them to solve safety problems quickly, effectively and efficiently. I was driven to find better ways of keeping everyone safe.

Each time I took the next step up, I was responsible for a bunch of precious humans and expensive projects where failure would be very costly indeed. At one stage, a government license to operate a million-dollar-a-day operation rested on me. We had to meet regulators' expectations before we could continue. To say I felt the fear is an understatement.

I notice the fear in people when I mentor them (although with the ideas gaining ground, the fear factor is reducing). One client talked about overcoming fear to address resistance issues with his CEO. He was rewarded with success, and you will be, too. Another client was nervous about pitching Safety II to her CEO. In conversation, she realised she had the resources (several years of study) to start the discussion.

I liken this to Joseph Campbell's hero's journey.[6] When I first heard about Safety II, I felt the call to adventure – and it's guaranteed to be one hell of an adventure! You will need a mentor. For me, it was the early support I received directly from Drew Rae.

But I certainly struck trials, tribulations and failure. Once or twice, I made a complete spectacle of myself. I've always found failing painful. I care about what others think about me and have had to learn to care less and maintain my focus on where I'm heading. The thing is, you must go through the process of failing forward to gain the necessary new skills.

The death and rebirth process (as Joseph Campbell describes it) is the hardest to go through. As a female lead in male-dominated workplaces, it usually amounts to understanding that I'm more capable than I think. Other people have their own holy grail moments. On the other side, I'm able to bring what I've learnt to others (this book is one way to do so). Then, the next call to adventure beckons.

THE HERO'S JOURNEY

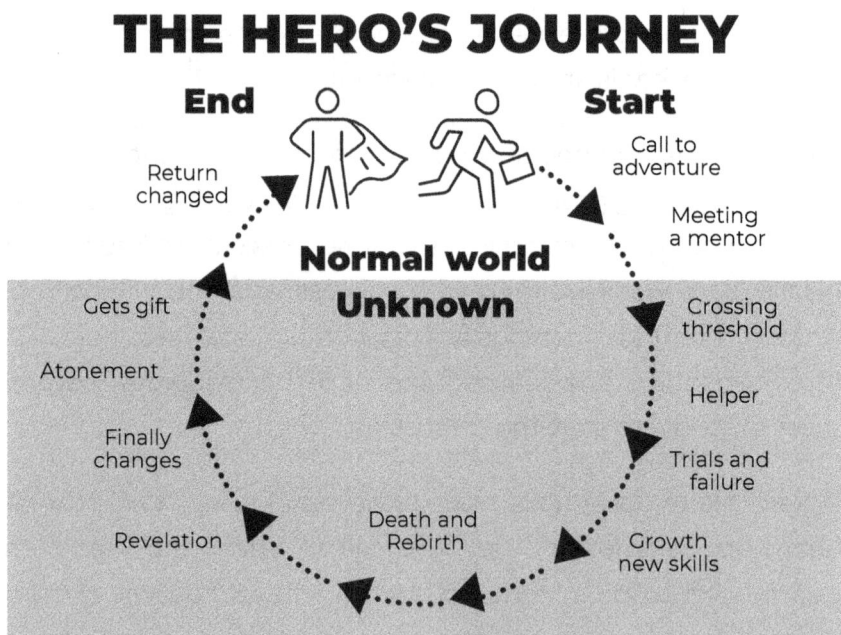

Figure 8: The hero's journey (after Joseph Campbell)

Building confidence and curiosity with leaders

How do you get your leadership on board with new thinking? The short answer is that you must persuade and inspire (obvious ingredients), provide comfort, and show confidence. That's because leaders need to experience how the new approaches work. It's the old adage of 'show, don't tell'.

Show your leaders how this awesome new stuff works. Keep things safe, easy and well-defined to assure your leadership and board. Enable a 'try before you buy' experience. Usually, that's an easy sell.

It's hard to let go of control and step through the unconscious fear this evokes. Your job as a safety lead is to co-create trials that demonstrate positive results to the leadership. These build comfort and confidence.

I've never seen this fail, and the reason is simple. Our teams are amazing people who rise to any challenge. They prove this stuff works. Your job is to set up the environment for them to do so.

Part of this process is the principle of 'failing forward' we've discussed, which is quite different to being a failure. Failing forward simply means trying something, working out what does and doesn't work and then adjusting, adapting and scaling up. Suck up as much of that as you can, and you'll be well on your way to success.

Mistakes are an inevitable part of this process, as are resistance, barriers, excuses, delays and politics. You name it;

I've encountered it. Make sure you meet the Readiness Test before you embark on any trials.

Running a trial, then planning a pilot

The most common approach to getting started with New View Safety is to start a mini experiment or trial. A series of these becomes a pilot in a more defined and structured area or process, which can be extrapolated into a rollout. Adapt thinking and implement in conjunction with your team. There is no one-size-fits-all with the experiments because the implementation of the principles should always be designed and adapted to your context.

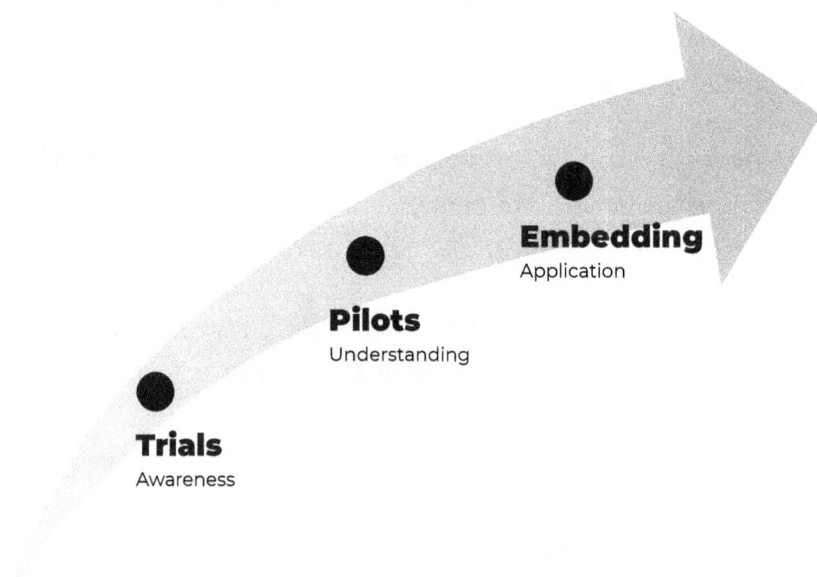

Embedding
Application

Pilots
Understanding

Trials
Awareness

Figure 9: Introducing New View methodology iteratively

The New Zealand horticulture industry wanted to trial Learning Teams in a culturally safe way with the RSE (recognised seasonal employer) workers who come from various Pacific Islands during the peak horticulture season as part of a government scheme. With a very limited budget, I designed a year-long project with minimal touchpoints across the industry to catalyse real change. The mental image I held was of skimming a stone across water, which always appears to defy physics.

First, we started with the smallest possible micro-experiment to exemplify the principles we were aiming for, get the steering team out of the mire of problems and challenges they could see, and focus on the opportunities at hand. To that end, we loosely organised a project that, as my first co-facilitator called it, we 'winged' the whole way. It needed to be small and easy to execute so we could quickly see the situation from a new perspective.

We didn't really design it; we knew what we wanted to achieve and leapt like frogs from lily pad to lily pad as we went. Within two weeks, we took management from being grumpy about the time investment with workers, to backing the project with a sales pitch to their boards of directors. An article on how the trial went was shared across industry media and was very well received. It was a vital component in driving the necessary change through the industry. Very little investment was needed to pull this off, and the effect was exponential.

The next challenge was to develop the concept into a short, structured pilot that covered the basics. Two organisations in different parts of the country were selected to participate, and a steering group plus a team of champions was formed to create and test the necessary elements. This project was conducted over three months with the appropriate engagement with all parties concerned. A professional video, including testimony from governmental stakeholders, has been produced to inform the industry of the potential of completely reframing safety.

Engineering innovation follows a similar process: start small with a proof of concept, then scale iteratively. A trial allows leadership to become aware of how the New View works. Experience shifts beliefs. Seeing it in practice, getting results and having a good experience drives momentum. Safe-to-fail experiments are low risk and designing a trial can be done on the back of an envelope. Another name for this process is an agile sprint.

At each stage in the horticulture project, we asked the participating organisations to share their learnings honestly with the next participant. This way, the spirit and intent were passed on to the next company, which could innovate the next stage. The horticulture industry in New Zealand is great at collaborating in this regard.

I often tell my clients to watch for the moment when their operations managers (or, even better, their work teams) start asking them to do more. Once they've seen the results of those doing the trials, they will want the benefits, too. When this happens, there is a genuine commitment to change. It seems to happen organically at the right time. Managers, teams and CEOs will ask for a Learning Team for a particular issue.

The prelude to this will be the effort you've put into engaging your teams in trialling the new ideas. If you choose the proper initiative, teams will take ownership of the process because they'll see its worth. Then, they'll drive the change themselves.

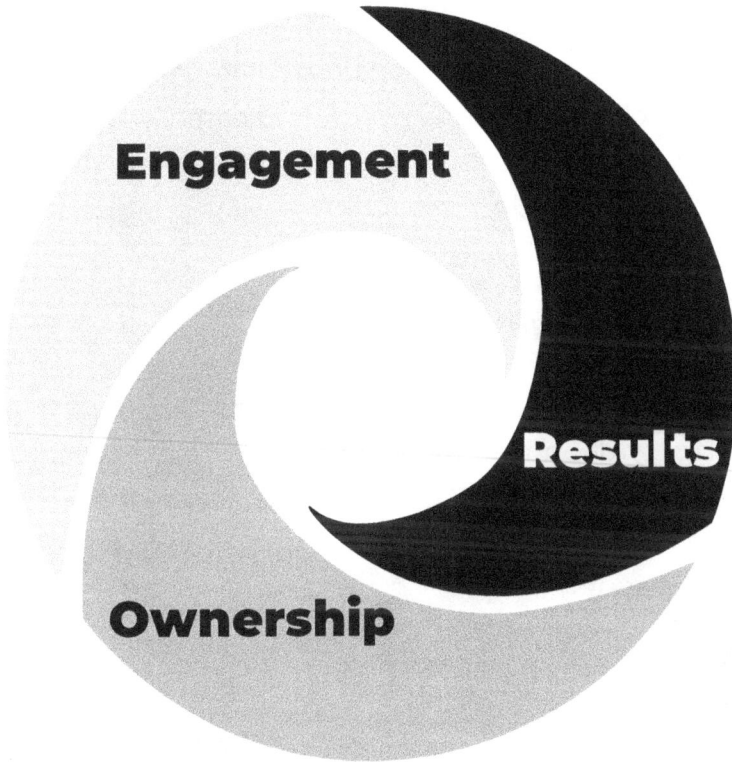

Figure 10: Tactics for driving momentum

The tipping point is when your organisation co-opts the process. It's the sweet spot you're looking for. After this, it's all about sustaining momentum. When you see dips in engagement, jump back into this and create some results to drive ownership again.

A word of caution. As an experienced Safety II coach, I see one of the risks in an implementation initiative is the failure to understand the importance of harnessing momentum. Think Newton's cradle. Getting an initiative going and dealing with the resistance to change requires investment. When the going gets hard, people sometimes give up. The initial effort has created energy, and it's only a matter of time before that connects with the real problem. Once the energy is released, you'll get push-through. Giving up too soon loses the investment. That can be hard to get back.

What is a pilot?

A pilot is an initial small-scale implementation used to prove the viability of an idea following trials or micro-experiments. The pilot project enables an organisation to manage the risk of a new idea and identify any deficiencies before substantial resources are committed.

Carrying out a pilot involves designing a basic strategy. There are six steps.

1. Set clear goals.
2. Define the scope.
3. Establish a realistic timeline.
4. Define the target audience.
5. Measure success.
6. Create a clear communication plan.

I worked with one company to discover how trust and transparency could be improved. We wanted to road-test the ideas and see how they were meaningful in an organisational context.

We formed a steering group of four people – small enough to create a 'container', which is Safety II parlance for a safe space for change agents to collaborate and make things happen.[7] We struggled with the scope of the project and had to engage with leadership and the board several times to ensure we had it right.

We ran the initiative after choosing the right people to be involved (work teams, management and directors). Clear outcomes were developed and communicated internally and with external stakeholders. A year or so later, we circled back to evaluate the project's success, creating a report for the funder to share findings and propose further research.

Working with the 'diffusion of innovation' curve

It's fair to say that not everyone takes naturally to the New View Safety concepts. And nor should it be so; everyone's skills and talents are needed in organisations. It's when we complement each other that great things happen.

The Diffusion of Innovation model (Figure 11), developed by Everett Rogers, shows that not all individuals or organisations will be ready simultaneously. Responding to this reality is a little like addiction theory. To make a change, the person must have contemplated the change and developed the will to go through the pain of it.

DIFFUSION OF INNOVATION MODEL

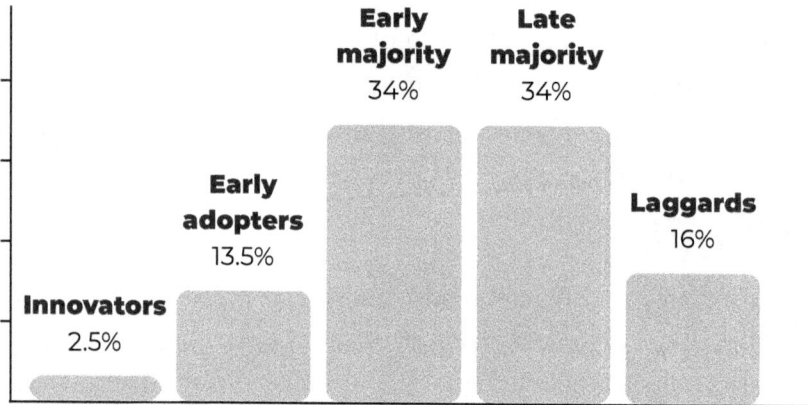

Figure 11: Diffusion of Innovation model[8]

One company chose a more mature team with a good operational leader to run a pilot, knowing they would likely engage with the ideas and produce results. Ownership by the rest of the organisation would follow when they saw a peer doing well with the methods.

Generally speaking, the rule of thumb is to start where things will be easier. 'Go where you're loved' is an important tip coined by Daniel Hummerdal.

One client wanted to start with the problem team because the pain was felt there, and results were needed. However, traction was slow due to a couple of managers who were bullies, and progress stalled when significant team members resigned. A much better strategy would have been to choose the team that was doing okay and could show the others what the approaches looked like in practice. That may seem counter-intuitive, but it works.

Getting all this going requires a powerful coalition. We're talking about a champion or two, or even a group, who will use their influence to carry the change. These people will often have an aligned passion and be willing to put their boots on and help deal with the resistance early on. I often find it's a progressive safety or HR lead, although it may be a CEO or head of operations (even a director). I've also had instances where I've partnered with the work teams themselves. Horses for courses.

I'm finding that younger people make the best coalition partners as they get into senior positions. Gen Y and Millennial generations want the change that Safety II promises. They jump in and get traction pretty quickly.

If you're a safety lead wanting to start the Safety II journey, find that coalition partner. They'll often be someone who can influence others in the organisation from a personal and positional perspective.

One client in a high-profile commercial construction company needed to get their CEO and the CEO of their major contractor on board. (More on contractor partnerships in Chapter Ten.) She chose a project director with inside knowledge who had the respect of both CEOs. With him in place, the right hustle happened behind the scenes, and a trial was designed and kicked off.

Using the right of veto

Establishing a constructed right of veto for your CEO (or whoever is sponsoring your project) is a great tool for catalysing momentum. Let me explain. When asking the CEO to embark on an experiment or pilot, a great way of establishing their support is giving them the right to refuse any initiatives suggested by teams that arise from the intervention. At times when I've forgotten to implement this, operations leads have suggested it, so it's clearly necessary.

The right of veto should be well understood by the team and accepted as a fundamental right of the CEO or whoever is sponsoring the project. When all parties understand this, innovation becomes freer. In this way, the business/operations lead gets to experience letting go of control with the comfort of knowing they can take it back whenever they want or need to.

One CEO I worked with was concerned that the process would unearth a whole lot of complaints and demands for more resources (a common fear). While he didn't explicitly say so, he didn't want to open a can of worms that would be difficult to manage. When handed the right of veto, he gave permission for the design of a trial. Once the team understood this right, they chose a starting point that worked for the CEO. Work teams always get what we're trying to achieve, and as long as everything is on the table, they will respect the boundaries and appreciate the honesty.

I usually communicate the right of veto during the launch of the initial engagement so the teams hear how it's framed to the leader. Sometimes, this is done at the beginning of an all-hands meeting when Safety II and Learning Teams are introduced. It could even be at the start of a Learning Team when a sponsor should come along and make a brief appearance to show support and an attitude of listening.

Whenever I've used the right of veto tactic, there are no problems. No one fusses, and the solutions proposed by the teams are constructive and well received.

With this right of veto in place, let's consider what specific ideas we are introducing to your leadership.

CHAPTER 6

Introducing New View Ideas to Your Leadership

Leading safety in the New View involves a shift in mindset

Key points

- Rather than tell and direct, seek to understand Work As Done.
- Vulnerability, empathy and trust develop high-performance relationships.
- Ensure your leaders are culturally competent.

A bi-monthly, all-hands meeting with staff included a reminder session on Safety II. It covered how the company was going with the New View Safety approaches and the challenges ahead.

What most excited me was that the session was written and delivered by the GM, not the safety lead.

After initially being sceptical about Safety II and worried about letting go of control, this general manager had experienced and truly embraced the process and was leading from the front. This is how Safety II *must* be implemented. Lead and owned by your operational leadership.

Too often, I see the safety team leading it and getting variable results. One leader told me recently that he constantly struggles with the discomfort of letting go of control, and his natural tendency is to have answers to every problem. However, the results had spoken for themselves, and he decided to step into the discomfort.

What he also did (but didn't verbalise) was embrace the vulnerability of giving his work teams a greater voice in how things were run. His operations team supported him in this and overall relationships and performance improved.

I subsequently had a conversation with the company director, who shared that since doing Safety II and Learning Teams, he can now see how smart and capable his whole team are, including the guys who do relatively unskilled work. He related a newfound respect for his people. The guy even gave me a hug – and he was someone I had regarded as a bit of a hard nut.

This all happened because the leader allowed transparency over what was happening at the work front and found collaborative solutions that embraced the skills and capabilities of everyone in the team. It was truly a win-win.

We can return to neuroscience to understand the impact of directors' mindsets on our organisations. The Pygmalion effect shows how expectations influence our performance. Sometimes called observer-expectancy theory, the expectations of people in authority influence those below them in a self-fulfilling prophecy.

The Safety Differently approach puts safety back in the hands of those at risk. It creates a positive Pygmalion effect with management treating the people at the sharp end with trust and respect for their skills and ability to make decisions, which, in turn, creates the kind of responsible behaviour we hope for in our cultures.

> Ultimately, the job of the safety lead is to set leadership and governance up for success with the shift in thinking and approach that Safety II requires.

If only I had a dollar for the number of times I've heard leadership say they believe the 'problem' is workers' behaviour. When things go wrong, they say the workers made dumb decisions. Executives will keep thinking like this until we provide evidence to the contrary.

When we make transparent the messy reality of what drives team capacity and how they handle variable demands, this will significantly shift how leaders observe teams and set an upward spiral where appropriate support is provided.

In this way, we evolve well beyond the obsession with compliance and having the workers behave in a certain way. When leaders and directors understand how the teams are creating success despite the imperfect systems they are providing, they start to see how smart, capable and switched on their teams are.

One CEO I worked with was happy to share his thoughts on a Safety II process we went through. He said, 'Looking at how this was done, and some of the outcomes, really focused us on altering what we were pushing down from the top to be compliant rather than anything else. I think that's a real lesson for me as a CEO. The feedback I was getting was good because it was real, even though it wasn't always what I wanted to hear. But it gave me a real insight into how we're performing, not just what our outcomes were. It was great, and I'd certainly recommend it to anyone.'

By the end of this book, you'll have a framework and reporting process in place to ensure this keeps happening. It will radically change your culture and enable trusting relationships as the positive impact of this stretches beyond improved safety.

So, what is the mindset shift needed, and how do we coach leadership using New View concepts, theory and tools?

Transparency and vulnerability to improve due diligence reporting

In 2022, I ran a trial with funding from our regulator to discover what transparency meant to an organisation. We ran sessions with the board, middle managers and field workers. The major learning was that transparency creates vulnerability for everyone. All team members, at all levels of an organisation, are imperfect, and no one enjoys feeling vulnerable. Everyone deserves respect for their hard work and best efforts. The board and CEOs are no exception.

Full transparency can lead to some pretty honest conversations. It's a bit like ripping off a bandaid – painful but necessary. What's on the table can be resolved and prevents all sorts of tip-toeing around situations that should be addressed. As we've discussed, this is one of the weak aspects of hierarchy. We found that teams generally feel respected when bad news is delivered honestly. Window dressing can make them feel undervalued.

The *Better Governance* report in Chapter One interviewed directors who said that, given their status and position, they felt pressured to have answers to problems. Saying they didn't know was challenging. At times, they would avoid asking questions rather than risk looking stupid. The concept of curiosity reframes this situation and gets us into the win-win zone.

Yet, to meet the legal requirements, directors must know and understand the conditions and constraints the teams are adapting to. That is the intent of the law, and it's how we hear the weak signals and prevent drift to failure. We must provide resources to support team resilience.

Our project had four findings, which started the journey to better reporting.

1. **Humanising senior company officers**

    For workers to honestly share the reality of how they are working without fear of reprisal, we need to deal with the power distance created by the hierarchical nature of our modern organisations. This involves humanising the people they need to be open with. During our project, the directors talked about their personal lives, families, hobbies and interests. Another option here is reverse mentoring, where younger staff members spend time with directors, sharing insights into their generational perspectives.

2. **Politicisation of information**

    Next, we must understand how the truth of how 'work is done' (more in the next section) flows up the food chain. Anyone who has ever been in management knows that some things are best dealt with at an operational level. We don't want undue interference when things should be expediently dealt with. However, pertinent insights also need to cut through the middle layer of management. Bad news is difficult to pass up the line. The CEO in the trial remarked that several themes were not made apparent to him.

3. Dealing with vulnerability

There's no easy answer to this except to discuss it openly. Work together as a senior team or board to find ways to handle it effectively. Later in this chapter, we look at practical ways to broach the challenges of building a culture of truth and honesty. This organisation arranged regular transparency sessions with the whole operational team and their CEO where any real questions could be asked with real answers given.

4. Link to capacity and demands

Instead of a constant stream of hazard identification and associated actions, focus on building team capacity and ability to absorb demands. In this case, examples were how emergency calls were handled to avoid fatigue, the ill effects of restructuring on procuring safety-critical equipment, and needing more input into training and competency expenditure from the perspective of on-the-ground expertise. Work teams suggested having more transparency over finances and budget allowances so they could give good advice to their managers based on what they knew were the priorities on a risk basis.

To really support teams well, as per the new definition of safety, we need directors to understand how they can better resource the teams. To make that a priority means the board requires a true picture of worker intelligence, warts and all. Senior leaders need to support this process.

One New Zealand organisation won an industry award in 2023 for dealing with this by having worker representatives meet regularly with a committee at governance/executive level. The effect was that middle management ensured the whole truth of how things were going reached governance before it was shared with the committee. Reps were also supported to design their own reporting and comms channels.

Chapter Twelve will cover the reporting structure. For now, what do we want transparency over?

Understanding work as done

In current safety management, leaders are keen to either *do* the right thing or *be seen to do* the right thing. That is laudable. Often, they go to lengths to seek assurance that safety protocols are in place and treat safety as a priority next to productivity.

Leaders, though, need to change the way they operate. They misguidedly think 'managing safety' is the best way to achieve safety. However, safety emerges from many factors, and leadership have less visibility over them than they think.

The New View Safety world has adequately captured this idea as *Work As Done* versus *Work As Imagined*.

An example is the best way to explain this.

A construction company gave written warnings to frontline staff after a similar incident occurred three times. Each time, they issued instructions to the workers to remind them of the protocols. The third time was a loud message, including videos, special meetings and other hyped-up communication methods.

It wasn't until the company realised something wasn't working that they tried a New View approach. After adopting a mindset of curiosity, the team soon answered the mystery behind the repeat events. It turned out that a Life Saving Rule was impossible to follow. The teams had to break it daily and weren't saying anything out of fear of reprisal.

This example clearly illustrates 'Work As Done'. Leadership thought they understood how work needed to be done (Work As Imagined) and kept hammering away at a message that was getting nowhere. Only when they humbly asked how work was really occurring could learning occur and appropriate solutions be found.

In Figure 12 below, the idea is that work never goes according to the plan in management's head (the black line). In reality, work is messy (the blue line). If we don't listen first and support the team to deal with this variability, we inadvertently expose them to the red line, where the teams get outside a reasonably safe zone and things can go wrong.

In the construction company example, the Life Saving rule (created by management) was the black line. The blue line was the day-to-day conditions the team faced where the rule worked sometimes but not in others. The fact that management failed to understand this and tried to create a better way to manage the situation meant they often ran into trouble and made do.

The red line depicts the worst-case scenario where the lack of open and honest communication causes the day-to-day reality to drift far from how managers think things are happening, meaning suitable solutions aren't arrived at. In safety science, we call this 'procedural drift'. When this process happens unchecked (which it often does), it easily leads to accidents.

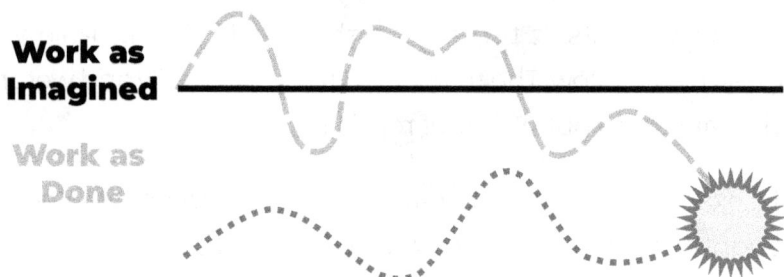

Figure 12: Work as done vs work as imagined

Adopting a mindset of curiosity can require sidestepping judgement, blame and ego — often when we're unaware that we're doing those things. As we've discussed, it also ultimately requires some vulnerability from leadership to admit there could have been a better approach and the humility to figure out how to listen more.

Depending on the situation, when we start with New View Safety, I often tell leaders that an 'inconvenient truth' will likely be revealed during most learning interventions. Otherwise, the problems would already be solved. Everyone wins when it comes out of the woodwork.

I did some fantastic work with a director of an industry body who wanted to improve safety and wellbeing outcomes for workers in a primary industry. We looked at how to improve cultural competency alongside New View Safety because many workers came from developing countries.

With the best of intentions, she insisted that workers needed to have better conversations with each other. I suggested we hold out the jury on that. After the initial experiment, management responded that they could now see how smart and intelligent these teams were and what great suggestions they had. Humbly, they realised it was management who needed to learn how to listen better.

I see this often. The bird's eye view from the top of the hierarchy, combined with the inherently political nature of our organisations, means that information about the reality of daily work is always skewed. That affects decision-making.

Through these approaches, leadership sees that the teams are handling day-to-day variability in ways that are unseen. Giving more trust to the teams does wonders for building their confidence. Trust allows the teams to develop more skills and capabilities and previously unresolved problems get sorted out.

Introducing New View ideas to your leadership

The starting point is exposure to the basic assumptions of New View Safety. Using new language helps leaders to wrap their heads around the required shift in thinking.

A small business owner in the residential construction industry wanted to use Safety Differently approaches, mainly out of frustration with mounting bureaucracy over the years. However, he realised that New View Safety is not just about less paperwork; it is a change in leadership style. He described the leadership challenges this way:

> 'How do I achieve this? Well, it's not easy. I'm old school and have a top-down approach to running the site. I give instructions and when they're not carried out to my satisfaction, I try to find a solution. In a Safety II environment, I'm supposed to engage with my guys and get them to help create the solution.

> 'From a safety perspective, that means I have to engage with my team and get them to speak up. But when they do, they want to please me and tell me what I want to hear. Especially the younger ones who may not know what a hazard is until it smacks them in the face or bites them on the arse!

> 'I'm facing the need to change my approach to allow and encourage an honest, open environment so that my guys will think about what they are doing and bring considerations to the table, no matter how stupid, without fear of being labelled an idiot. Hopefully, in the course of time, they will bring constructive and productive ideas that I would not have thought of.'

The answer to my colleague's dilemma is to develop coaching skills. Coaching draws on curious questions and effective listening to build shared understanding and awareness, increase responsibility and take ownership of decisions and actions. Coaching conversations explore what is currently happening and all the possible ideas and solutions a team could adopt.

The following principles provide an ideal starting point.

The five principles of human performance

Teaching the HOP principles is a good starting place when working with leadership transitioning to the New View Safety thinking. Now widely known, HOP stands for Human and Organisational Performance. It is a set of assumptions that leads to new perspectives on how to view and respond to worker mistakes.

We see improved performance and safety when this mindset is embedded into a culture. That's because people feel safe to be transparent about mistakes, improvements are made and trust develops. It can take time and effort to embed this thinking, but it's a real game changer.

Let's briefly look at each of the assumptions:

1. **Error is normal. Even the best people make mistakes.**

    The first step is to fully understand the science around mistakes. Neuroscience and psychology clearly show that making mistakes is normal. Humans need to adapt, create and innovate, so mistakes guide the evolution of our species. Our brain physiology lends itself to developing efficiency over time, meaning there is a trade-off with occasional errors. We'll cover more of this in relation to error traps and performance modes in Chapter Nine.

2. **Blame fixes nothing.**

    Blame is also a human response that leads to fundamental attribution error, where we assign qualities to a person based on an observed behaviour. Power imbalance makes it easy for leaders to paint a worker as 'dumb' or 'stupid' when something goes wrong. As Todd Conklin puts it, the problem with this is: 'You can't blame and punish at the same time as learning and improving'.

3. **Learning and improving are vital. Learning is deliberate.**

    If you've agreed with the assumptions listed above, the next landing point is to set aside conclusions about what went wrong in situations and seek to understand from the worker's point of view. We are guaranteed to get a completely different set of data. Often, there will be an inconvenient truth where leadership discovers something they were previously unaware of.

4. **Context influences behaviour. Systems drive outcomes.**

    The context of organisation systems leads to certain behaviours. The most common influencer is production pressure which is purely in the domain of the managers and organisation. Senior, experienced team members can stand their ground regarding the pressure from production requirements, especially in a healthy culture. However, most struggle to find constructive ways to adjust to how the job is done, even though a two-way conversation would usually result in win-win solutions.

Teams often successfully handle the pressure regardless, but it sets people up to make mistakes.

5. **How you respond to failure matters. How leaders act and respond counts.**

   Trust builds when you demonstrate that judgement is suspended and collaborative learning is sought in a failure situation. This leads to more openness when issues occur and better performance overall. Be wary of using discipline when understanding and learning would be a better response for long-term results.

To implement HOP, the first thing to do is to ensure your leaders understand the fundamental ideas and what they mean in practice. Coaching your leaders involves teaching them to remain curious, employ empathy and build trust when having conversations with teams. Their job is to deliberately set aside their answers and remain open to others' solutions and perspectives.

One team shared with a leader that they take some safety shortcuts when there is too much pressure to meet unrealistic deadlines. Trade-offs become necessary. The leader wanted to blame the workers for not strictly adhering to safety procedures. The challenge is to listen without judgement so more effective solutions can emerge. Telling people off forces the problems back under the surface.

There is nowhere this comes into play more than how teams deal with production pressure. The impact of this comes up every time I support organisations in getting started with Safety II methodology.

I've found that managers tend to have a fixed viewpoint on issues surrounding production pressure, which generally amounts to some polite (or not-so-polite) version of 'too bad, suck it up'. Understandably, this relates to the central KPI of making a profit. However, remaining curious almost always elicits the team's views on how production pressure is often causing inefficiencies that are easily solved with everyone's input.

I have countless examples of this. One group of workers told me their managers were constantly on their backs to speed up production. As the workers didn't want to complain too much, they had to make quality sacrifices to keep up. They weren't being lazy; it was simply the best they could do.

During the sessions, I facilitated a deeper understanding of the situation, including how performance and quality could be improved while supporting the team's wellbeing. Great insights and suggestions were put forward. These included balancing the mix of faster and slower crew members in different crews, and creating clear communication with managers who lead production and quality and sometimes give conflicting messages.

One of the best ideas was to support the teams in meeting the KPIs by installing electronic boards that kept the team updated on how much output remained for each day. That way, they could support management with the 'how' of achieving daily targets by using their initiative.

These sorts of insights are readily available from your work teams every day. The answer lies in getting out and having conversations.

Shifting the focus in leadership walkarounds

One way to begin the shift promised by the HOP principles is to change how leaders carry out their safety walkarounds. This is a common safety management activity in modern corporates where leaders and directors visit workers and have conversations.

They should focus on discovering and analysing *what is helping and hindering performance through the views of different roles and levels across the organisation*. Establish a bank of appropriate suggested questions and ask three or four per month. Rotating them over a year will keep things fresh. NZ's Business Leaders' Health & Safety Forum has a useful resource to support leaders. https://www.forum.org.nz/assets/Uploads/Guides/Learning-from-Success-2017.pdf

These activities with leaders – teaching them the HOP principles and refreshing the walkaround conversations – will set them up for using Learning Teams. We'll cover that in the next few chapters, but first, let's discuss the all-important topic of how to build a culture of telling the truth to support the needed transparency over system vulnerabilities.

The truth will set you free

A regular client asked me to support them with a Learning Review (see Chapter Nine). After a fairly serious incident, they were worried that many people had known about the troubles that caused it, but company systems and culture weren't creating the necessary visibility at executive level.

Through the discovery process, it became clear that, among other things, some middle managers felt the pressure to create mostly positive reports to the managers above, so there wasn't full transparency about what was happening. This is normal in organisations that are driven by hierarchical systems. We've all been in this situation. Middle managers aren't the problem. The problem is what is aptly termed an ethical dilemma.

> An ethical dilemma is where two ethical outcomes come into conflict. Usually, in the safety world, it relates to the need to make a profit and pay the staff (rather than go out of business) and the need to do the work safely.

In the example, an appropriate company strategy of investing in technology to support the future environmental needs of society was being prioritised. A message of cost-cutting in other business areas had resulted in organisation-wide misunderstandings of the true boundaries and how to push back if safety margins were being eroded.

Everyone was doing their best.

One way to resolve ethical dilemmas is to build a culture of always telling the truth so we get cut through when needed. And that's not easy in a Western culture where subtle lies prevail for the sake of self-preservation. Innocent politics and other normalised deviations from the whole truth are accepted as how things are.

To the total credit of the senior executives, they quickly concluded that they needed to be more truthful and transparent with their staff about the strategy and its implications. To this end, an all-hands meeting was held to provide open and transparent insights behind strategy and budgeting policy, and a current review of how work teams should be involved in budget and expenditure is underway. Money was spent to fix the immediate issues.

These approaches require a mature management culture. Not perfect, mind you – no leadership team is. None of us are. While resistance emerged, the organisation was willing to look at new approaches. In this case, they were ready for change.

Another essential area of change is focusing on how we work with team members from other cultures.

Building cultural competency as a leader

One organisation I worked with had made great strides with Safety II. They were buoyed up and regularly tried new ideas with great success. However, whenever I went to their all-hands team meetings, I noticed how the stereotypical white, middle/upper class, mature and educated management spoke to the mainly Pacific and Māori work teams. Management were pleasant, but they did all the talking. It wasn't a two-way conversation.

Cultural competence is the ability to understand and interact effectively with people from other cultures. To have multicultural competence, you need the willingness to learn about others' cultural practices and world views. It requires a positive attitude toward cultural differences and a readiness to accept and respect those differences.

Māori, the indigenous people of New Zealand, comprise 15 to 20% of our society. They have a rich and proud culture and heritage. A presence across all levels of leadership and the workforce, many are proudly working class. Smart, capable and supportive of their teammates, Māori tend to be honest and insightful about the work. As informal leaders in their work teams, they have proportionately more to contribute.

In New Zealand workplaces, like many OECD countries, immigration policies mean we have a range of ethnicities working together. People from nine different Pacific Islands work alongside immigrants from China, India, the Philippines, Indonesia, Japan, South Africa, the UK and many more.

While the dominant culture in New Zealand is currently European, fostering trust between leaders and workers requires learning about the various cultures within our workplaces. In this way, we can understand others' cultural values and how these play out in how we work together effectively. Over time, we can expect a shift as New Zealand's culture evolves.

The New Zealand horticulture industry hires many people from the Pacific Islands and other ethnicities. The sector is meeting this challenge head-on through cultural competency education alongside Safety II trials. Designed and delivered by the Ministry of Pacific Peoples, with a Pacific facilitator, a deeper understanding of the Pacific culture and drivers is discussed. This enables empathy and understanding from horticulture managers who say the sessions are extremely valuable.

A general introduction to the Pacific Islands and the cultural values is followed by discussion about why better engagement is worthwhile. Tools and frameworks provide methods. Tips and tricks are passed on and common situations and scenarios are discussed. Managers can see some of the reasons they have had difficulty communicating before.

The concept of 'Vā' resonates with management. A Pacific word, it roughly translates as 'creating a space for long-term relationships based on mutual trust and co-operation'. It gives management a better understanding of how to relate to workers and what would help create smoother work for everyone.

The management of one kiwifruit company found that, following a learning process, the work teams from Samoa had some great ideas to contribute to improving the production process. The team manage their farms and carry out trades back in Samoa. While they performed relatively unskilled tasks in the packhouses, they easily had the acumen to contribute to process improvement. After using the cultural competency ideas, they were invited to contribute to the plans for the next season.

Another grower engaged with their team from Vanuatu and found they could readily point to ways the company could work collaboratively to smooth out production issues. They explained that, culturally, nominating a team member as a 'go to' conduit for conversations back and forth to management would supply ideas and responses to feedback. In the evenings, as they would have done in the village, the team got together to discuss matters and feed the information back.

When I presented the results of these trials to the industry CEOs and board chairs, they overwhelmingly saw the potential of the approach to transform their industry.

With leadership on board, let's get into the mechanics of understanding the all-important conditions and constraints our team are navigating.

CHAPTER 7

Introduction to Learning Teams

Getting your New View journey started

Key points

- The core Learning Team explores normal work and seeks to understand the adaption to variability required to complete the job successfully. Listen to your teams; they are the experts.

- Shift from a reactive culture to a high-performance culture by understanding what drives capacity and how your teams handle demands. Explore the conditions and constraints they are adapting to.

Learning Teams introduce the Safety Differently concepts in a structured, controlled manner. They lay the groundwork for changing systems and structures, which we'll cover in the next book, by giving transparency over system vulnerabilities. When leaders respond well, trust is developed. For now, let's consider how to get started with discrete, facilitated, workshop-style activities. Once embedded, your organisation will fully understand how the principles operate, enabling further systems innovation.

> Learning Teams are structured to facilitate team-based enquiry about normal, everyday work. The aim is to understand what contributes to successful work, how this can be leveraged, what makes work difficult and how this can be addressed positively.

The learning centres on discovering and analysing what is helping and hindering performance through the views of different roles and levels across the organisation. Learning Teams help us understand how work really gets done — not just how we think it does.

In this way, we understand the conditions and constraints that create variability in day-to-day work, and which the teams successfully and safely adapt to. With this New View of safety, we can focus our improvements around these areas, rather than introducing more rules and paperwork. That is how we improve team capacity and resilience.

Learning Teams should always have a proactive focus. In the safety world, we're so used to investigations when something goes wrong that people assume a Learning Team is just another form of the same thing. That is one possible use for the process, which should be forward-looking (which we'll cover in Chapter Nine). However, the sessions should happen long before anything goes wrong. In Chapter Eleven, we'll look at a framework to ensure proactive learning is robust, dynamic and consistent enough to get results over time.

As the diagram in Figure 13 shows, we mainly want to learn from situations where work is difficult or successful. Of course, this is most of what happens in workplaces every day. Learning when things go wrong is only a small part of the picture because incidents happen rarely in relation to the number of hours we work.

FOCUS OF LEARNING

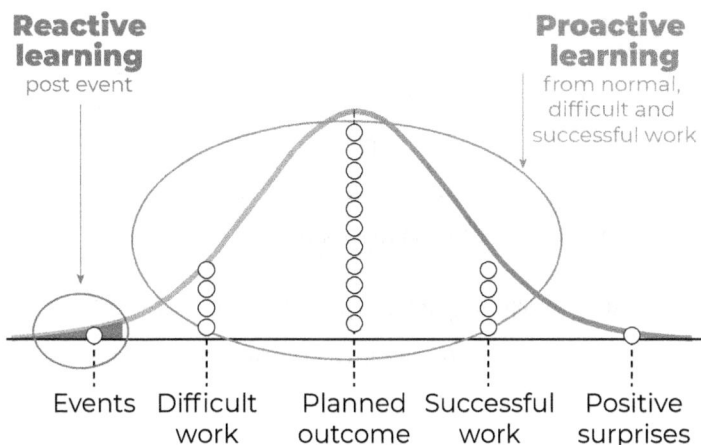

Reactive learning
post event

Proactive learning
from normal, difficult and successful work

Events   Difficult work   Planned outcome   Successful work   Positive surprises

Figure 13: The focus of learning on a wide spectrum of work outcomes

A Learning Teams approach moves us from blaming teams for work difficulties to learning how they handle difficult work successfully and how improvements could be made. That leads to insight and innovation, engagement and operational resilience. A Learning Team is a continuous improvement process to manage operational activities and risks. Most importantly, the people who do the work are part of making it better.

In addition, we rarely stop to learn when work has gone well and should take the time to do so. The insights will be even more valuable.

Learning Teams are almost always a good experience for everyone. Management gains great insights into how to solve challenges *with* the team, positively affecting working relationships. The improved trust and collaborative mindset flow into day-to-day work. Done well, Learning Teams can massively improve your working culture and ability to harness everyone's capability.

After one Learning Team session, a supervisor remarked that the team instantly became more confident at sharing their insights with him during the day and showed new leadership with problem identification and solving.

In this chapter, we'll cover the overall methodology and, in the next, we'll address how to facilitate Learning Teams.

The power of harnessing team-based learning

In 2015, I did an early version of what is now called Learning Teams. I was asked to do a board paper for an industry body in the agriculture sector. I brought team members from several farms together and ran two separate workshops. The first was a discovery process around what was going well regarding risk and their frustrations with it.

A month or so later, with collated results and some analysis from site visits, we conducted the sessions again and asked for their solutions. We then put budget allowances against them, and I wrote up a formal paper making it clear that the central methodology assumed the team members were the experts in the risk.

The board loved the paper and signed an open cheque for all the recommendations. I was stunned. I'd never had such good results. However, the board was sceptical of my expertise. They asked, 'How do we know some lady from the construction industry knows anything about our risks?' (No kidding, they actually asked that.)

They didn't get the part about their teams being the experts (we've come a long way in the last decade), so they hired two farming consultants to peer review my paper. They declared that my findings were virtually perfect. My first New View experiment proved unequivocally how capable the teams were.

I have seen phenomenal results with just a single Learning Team, from hundreds of thousands of dollars saved to turning around an entire culture or completely reinventing an operation. Sure, not every Learning Team has this result. Yet when organisations remain consistent over time, operations improve, staff are happier and previously unresolvable issues magic themselves away.

Much has been written elsewhere about Learning Teams, and this book assumes a basic knowledge of their concept, purpose and process. I'd like to deal with attribution here. Todd Conklin is well-known to many in the safety world. He originally conceived Learning Teams as a discrete way to introduce New View Safety concepts.

Todd and many others who have since collaborated with him to develop the ideas and drive global change, are true pioneers. Without their dedication to change, we would not have progressed as much in the last decade.

The first step is to introduce organisations to the idea of proactively learning with the teams so we can support them to work safely, reliably and productively given variable conditions. We want to create positive adaption over time rather than allow them to drift into failure by listening deeply to our teams and supporting them adequately.

Learning Teams can bring about transformative change. Workers tend to feel heard. We gain transparency over system vulnerabilities. The team offer ideas on how to improve informed decision-making. Over time, trust and confidence build and relationships improve.

Establishing the need for a Learning Team

In simplest terms, a Learning Team aims to solve problems, but it has a very different view of what the problem is. The starting point is identifying a topic – usually a review of an area of work. Note that we don't usually choose a safety problem per se.

The best way to establish the need for a Learning Team is with the operational leads. Ask them what work issues are going on and what difficult areas are worth learning about. Here are a few examples I've been involved in:

- A project analysis following repeat incidents with key workers.
- The review of a training centre with the students.
- A contractor review with the regular contractors involved.
- A repeat failure of machinery or plant.
- Follow up to a serious near-miss incident.
- Budget overruns and difficulties on the job.
- A project with no real problems, delivered on time and budget.
- Reviewing a high-risk operation with traffic management and forklift use.

In choosing the topic, you and the team might explore a stubborn production problem, repeat incidents, or another work area where there's value to be gained. You might also choose an area where everything seems successful and understand why so you can share learnings with others.

Sometimes, a Learning Team isn't necessary, or problems emerge that don't need to be solved by the group. Technical problems should still be resolved via normal means within organisational processes. Adaptive problems (see Table 3) are the perfect fodder for a Learning Team. But if something can be reasonably easily solved via usual channels (e.g., a technological improvement to some equipment), don't bother.

If it's an issue you've tried to solve with technical fixes and is still a concern, it's likely to be an adaptive challenge.

Technical problems	Adaptive challenges
Easy to identify	Easy to deny
Quick and easy solutions	Require change in beliefs, roles, relationships
Solved by authority or expert	Best solved by people with the problem
Require change in one place	Require change in numerous places
People are receptive to solutions	People resist acknowledging challenges
Solutions implemented quickly	Solutions require experiments and new discoveries

Table 3: When to use Learning Teams

Understanding the real problem

The biggest trap I see is when people select safety problems as the topic. For example, they decide to look at why they are having a lot of manual handling injuries, which is a Safety I focus where you are trying to avoid things going wrong. A better approach is to examine the work area where the injuries occur and determine what conditions and constraints are inherent in the work. Several factors are likely placing unnecessary pressure on the team and creating goal conflicts. When we solve these, often the manual handling problems are inadvertently solved.

'What problem are we solving?' is the most common question from operational leads at the beginning of a Learning Team process. This question is indicative of the entire point of running a Learning Team. If we knew what problem needed solving, we'd have solved it already.

It also alludes to the fact that operational leads tend to be selected for their skill and being relatively independent in solving problems. Adaptive problems are best solved collectively and as close to the problem as possible. That means the problem needs to be *discovered*. We need to start by seeking to understand. It requires suspending judgement, patience and an ear to listen. From that process, we understand the real problems. It's as simple as following the process illustrated in Figure 14. Once you reach the end, start again until you've embraced a new way of working together.

THE DOUBLE DIAMOND

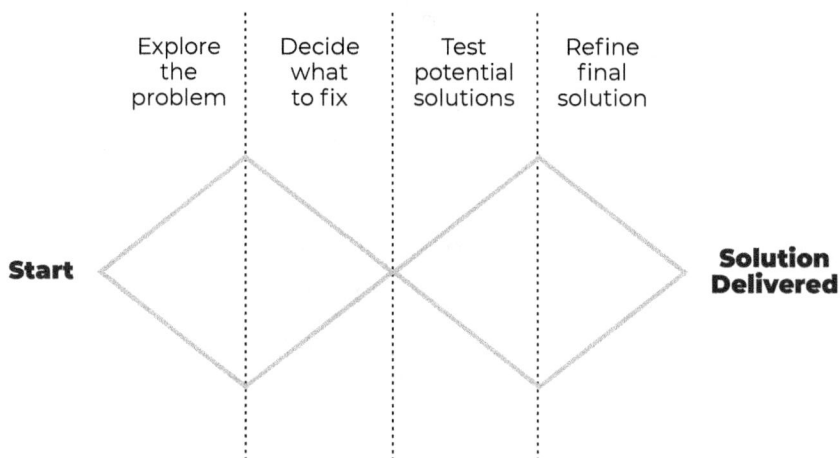

| Explore the problem | Decide what to fix | Test potential solutions | Refine final solution |

Start

Solution Delivered

Figure 14: The double diamond – discovering the problem

A great example of this important point happened a few years back when I was starting out with Learning Teams. Two competing companies asked me to help them with a Learning Team based on the exact same topic. It was an uncanny coincidence, so I was keen to see how the results would compare. Naturally, all information about each company remained completely confidential. My role was purely to ask questions and facilitate their solutions to the problems.

As it turned out, they started with the same topic and the same difficulties, but the problems identified were quite different. It highlighted that while technical fixes can be predictable, adaptive challenges depend very much on context. In this case, one company struggled with resourcing during COVID lockdowns. The other had issues with a client being too controlling of protocols and requirements and limiting the adaptability of the teams.

Letting the problem emerge is more important than you might think. Think of the O-ring seal problem when the Challenger shuttle blew up in 1986. The real problem is often known, but political or financial issues prevent commonsense from coming to bear. I call it the 'inconvenient truth'. The problems that emerge from Learning Teams bear these qualities and everyone wonders why we didn't take the easy road and solve them earlier.

The Learning Teams process

Attending a comprehensive Learning Teams training course will step you through running a Learning Team and provide the required skills. You can visit my website for course information; however, Figure 15 illustrates the basic ideas.

The process involves discovery, analysis and improvement phases.

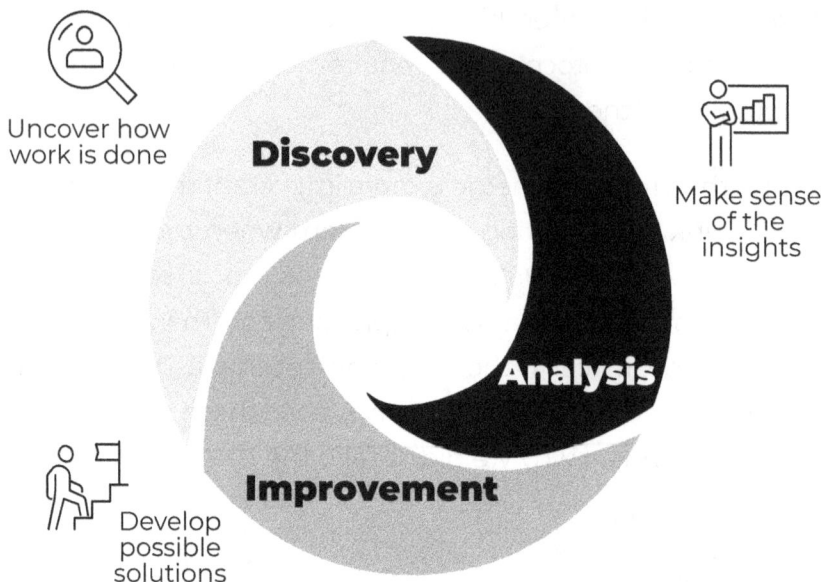

Uncover how
work is done

Discovery

Make sense
of the
insights

Analysis

Develop
possible
solutions

Improvement

Figure 15: The basic process during a Learning Team

Let's use a case study to walk through the Learning Team process. Applying these concepts in a real-life context will bring them to life.

Across New Zealand, the horticulture industry brings in extensive national and international revenue. Staffed by an array of seasonal and permanent workers, the peak seasons of picking and packing fruits and vegetables are hectic and require long hours. As mentioned earlier, teams come from the Pacific Islands in a government-run scheme. Backpackers join in to earn a quick buck, and a staple of Kiwi workers join the mix.

The sector wanted to trial Learning Teams. We decided to run an initial trial reviewing a busy kiwifruit packhouse. It was nearly the end of the peak season and the findings would support the preparation for the next season.

What we wanted to discover

The teams need to perform their jobs safely. For the most part, this means they are well-resourced and have the means to secure the tools, equipment and other requirements to do a good job safely.

To understand what conditions and constraints they were dealing with, we had to know the messy details of how they did the work. We were keen to know what would surprise management and their strategies to get work done despite variation, dependencies and difficulties.

Appreciative inquiry is a way to engage groups of people in self-determined change. It focuses on what's *working* rather than what's *not*, leading to people co-designing their future.

At its core, Learning Teams is about asking better questions. That moves us away from control, which is essentially telling people what to do.

Here is a core list of great questions to ask the teams.

- What makes work successful/difficult?
- How is success created?
- What helps and hinders performance?
- What tools, resources and strategies do people rely on to achieve success?

- What conditions and constraints make work difficult?
- What mechanisms are in place to understand how success is created?

Workers are smart. If you present this set of questions at the beginning of the session, they get what you're driving at and the conversation begins. Go through each question to progress the discussion based on outcomes that support better performance and adaptability and, in turn, ownership and co-leadership of the sessions.

Drawing out themes to understand capacities and constraints

In the packhouse example, my co-facilitator and I used question sets to understand the work and elicited a bunch of insights from the team. From there, we separated the responses into two lists: one with insights into what *helped* and the other with what *hindered* team performance.

From these insights, we drew out themes that described the necessary capacities to help performance, detailing areas where things that currently hindered performance could be improved. The conditions and constraints identified were:

Cultural relationships

The Pasifika teams wanted good rapport with the company to ensure relations with the villages in the islands meant a continual flow of workers over the years. This need influenced whether they would speak up if there were issues. The better rapport they felt with the company and management, the more they felt able to speak up, linking to the concept of Vā, as explained in the last chapter. A friendly, kind, mutually respectful relationship was important for effective work.

Production pressure

The nature of seasonal pressures and constraints meant production pressure was inherent in the workflow and drove requirements, which fluctuated throughout the day.

The extent of manual handling required, combined with production pressure, caused the teams to lift more than was recommended in an attempt to do the right thing by their employers. This was linked to the first condition. In their minds, they were contributing to the relationship by going the extra mile.

Training support

With the seasonal nature of the work and itinerant workforce, there were barriers to providing adequate training support.

Procedures and instructions needed to be more accessible and user-focused, e.g., language translations and training on the use of technology and devices.

In addition, it was apparent that the teams had skills from previous employment and businesses back in the islands that could assist with problem-solving in the packhouse facility.

Tools and equipment

Tools and equipment were important to good work and generally considered adequate and appropriate.

Wellbeing and fitness for work

The combination of production pressure and the challenges with training support, meant wellbeing and fitness for work were essential to support work performance, especially in a foreign culture with climate differences. The 'zero to hundred' nature of moving from hot climates with a slower pace of life to the colder, faster-paced New Zealand climate can lead to illness and injury early in the work placement.

While none of these were big surprises, the packhouse management were now armed with a new understanding of what drove capacity and resilience in their teams. They were very engaged and expressed a deep desire to continue this method of working with their teams. After a 'soak time' where the workers could consider if any further intelligence was worth sharing (overnight), we did a further session to develop ideas for improvements.

In this case study, the safety manager remarked that it was apparent that the real reason too many boxes were being lifted was different from what the managers thought. It was assumed that the workers were trying to outdo each other, but the process showed that the teams were simply trying to keep up with production, as a new artificial intelligence upgrade in the processing facility was causing teething problems. Of their own initiative, the work teams had created a flexi-team solution to absorb the production issues. They also took it upon themselves to lift extra to avoid needing more manpower.

The operations manager also shared that, following the sessions, he better understood the workers' individual strengths. He recognised untapped skills that were unknown before the learning process. Management planned to engage differently with the teams going forward.

Let's summarise the main findings here concerning Safety II. The teams were improvising a flexi-team solution to the production issues. They were creating success via adaptation to conditions (processing line issues). Smoothing out these processes improved safety and supported the team's ability to adapt successfully.

The safety lead remarked that he'd previously been unaware of the intelligent input the teams were capable of. By his own admission, he realised he needed to engage with them for answers that could be better than his own. This is a common outcome of the Learning Teams approach.

Analysis and ideation

When facilitating a second session, I asked the work team to imagine themselves as owners of the company. What would they recommend to themselves if they were responsible for the good running of the organisation?

Generally, you won't get fully formed solutions in early sessions; you'll need to follow up the Learning Team with more defined sessions and develop strategies that require testing and iterating. Doing this with the teams retains and builds the trust created by the learning sessions.

INNOVATION PROCESS

Figure 16: Stanford Design School innovation process

Once again, use concepts that are commonly used in engineering development. First, understand the needs of the user and define their problem. Next, devise potential solutions with the aim of prototyping and testing ideas. The Stanford Design School model (Figure 16) is excellent if you want to research this further. Note that you are continuously circling back to empathise with the team as the foundation of the process.

I like to use the Agile sprints approach to devise a concept quickly, build only the necessary elements to show the idea to others, and then get feedback and user data. Designing alongside the end user can be time-consuming, but the result is far more successful.

In the case study, the following three challenges were identified along with solutions:

1. **Production pressure and variability led to inadequate manual handling practices.**

   Solutions focused on improving the flexibility of manpower and two-way planning meetings with the team so they had more support for handling production pressure and variability. It was suggested that one of the Samoan workers should be given a supervisor role to provide more direct support to the teams. (Far better than more posters and lifting technique reminders.)

2. **Improvement of training practices given the seasonal nature of the work combined with language difficulties.**

   A more in-depth Learning Team on this topic was suggested to understand the complexities better. The use of videos and signage in a variety of languages was considered.

3. **Collaborative review of tools, equipment and work improvements regularly to maintain good resourcing.**

   A regular cadence of short learning sessions, including representatives from various teams and crews, was suggested to get real-time feedback on these issues. Pacific liaison officers, provided by the government, or trusted in-house interpreters were suggested as a resource to bridge the language barriers. Learning Team questions were translated into the dominant language of the work teams.

A report was written to explain the detail of the recommendations, and tabled with the board of directors. Having done many of these, I know that everyone walks away with a smile. The teams know they have been heard and there is better understanding together. Management is usually enthusiastic about doing it more regularly. In this instance, the CEO soon asked for more Learning Teams and had great ideas about the next topic.

Using artificial intelligence (AI) to drive your Learning Team process

As I finalise this book in mid-2024, using artificial intelligence in contexts like running Learning Teams is a hot topic. Using multi-language models, AI mimics the human intelligence required to analyse qualitative data.

From personal experience, using AI during a Learning Team is quite exciting. It speeds up the process and enables a more polished output. I think of using AI as a turbo boost – or facilitating a Learning Team with a jetpack on.

The potential of AI is to heighten our ability to gather and analyse collective intelligence in organisations. The theme of this book is to shift from leaning solely on the smarts of managers to including the smarts of workers. There's a lot of information to process and future AI will support us on this journey.

During the Learning Teams, we can use AI to:

- process the discovery work using AI
- analyse feedback in less time
- summarise findings and generate themes
- generate reports on the fly.

The trick to using AI well is understanding the three stages of the process shown in this table.

Safety II outcome	Using prompts	Validation
Output should be a clear understanding of how teams are adapting to conditions and constraints.	Select simple, clear, direct prompts that drive towards the outcome.	Check all outputs and redraft using critical thinking, empathy and insight to support the development of trust and ownership in teams.

Table 4: Three steps to using AI to power your Learning Team

The starting point is to agree that you want an output that describes the team's work based on the Safety II definition. In the next chapter, we'll look at essential data processing prompts as we go through the analysis process.

With AI, you must validate the output. In lay terms, that means rewriting the output using critical thinking to ensure it achieves the desired output. Frame the findings within the Safety II definition and purpose.

In practice, current AI can't and won't capture the empathy required to produce good Learning Team findings. Remember, we want to build trust between our work teams and their managers; only humans can do this. AI can be used to produce base output, and we must develop and redesign it to achieve that end.

To start using AI, tinker around and try out different combinations of tools. Experiment, practice and get creative. Before you do, though, be sure to review the policies and procedures for using AI at your workplace.

If you're in charge of facilitating New View Safety sessions, the next thing to understand is how you will go about it.

CHAPTER 8

Strategies for Facilitation

Building the skills to enable trust with your teams

Key points

- The secret to successful facilitation is creating an emergent space to enable change.
- Taking time to construct trust is the make-or-break point.
- Follow a tried-and-true facilitation process to achieve success.

One company in a high-risk sector is well respected across New Zealand for working to adopt Learning Teams before anyone else. Their strategy was to train a team of facilitators across the business to enable the resourcing behind the commitment to Learning Teams.

Time and time again, when I've worked with organisations that are introducing Learning Teams, success or failure is down to two things. The first is the readiness of leadership to embrace new ways of working together, and the second is the ability to facilitate well.

Developing a good facilitation strategy for your organisation depends on size, existing capability, and practical considerations such as accessibility. One size does not fit all. Hold regular training to bring new facilitators up to speed so the burden doesn't just rest on a few.

To put it strongly, a facilitator has to create an environment that shares power and listens to what's driving outcomes. They must be able to build trust between management and work teams when there is resistance, mistrust and even anger. Sometimes, that means putting aside alliances. They also need to understand how Safety II works in practice.

I often come across people who think they will be pretty good with this stuff and then struggle to get the right output from the Learning Team. Or, more often, companies start doing Learning Teams simply as a collaboration session rather than implementing Safety II in its full definition.

I want to state clearly that these are not simple collaborative problem-solving sessions, despite first glance. You need to draw out of the team how they currently create success in varying conditions before you look at solving any problems. This requires practice and skill.

Here's an example. A group of contractors formed a Learning Team at their client's request. It turned out that one of the client's competitors had successfully dealt with the issues at hand. Now, when it comes to safety, there is no competition. We're all in it together. Looking for success is an art to be developed and can involve looking over the fence!

The best way to learn how to do this is via coaching from someone who has demonstrated skill in this arena. To that end, training, mentoring and assessment are important to build competency across your organisation for an effective Learning Teams programme.

A word about new Learning Team facilitators: many safety leads feel nervous about running their first Learning Team. This is normal. Fear of failure is real. Allow them to learn and grow without pressure. Mistakes will happen; it's the only way to learn. The best thing is to jump in and have a go. Expect them to do a few Learning Teams before they get the hang of it.

I love how Josh Bryant, an admired Australian Safety Differently practitioner, calls it 'bumbling your way through' till you build confidence. People are often reluctant, but there's no other way. Like learning to ride a bike, you'll be good enough in no time, and the skinned knees will be forgotten.

Let's turn now to the skills and characteristics needed of facilitators.

Characteristics of New View facilitators

The reality is that not everyone is equal when it comes to the ability to facilitate well. A rule of thumb is to go with those team members who gravitate towards the task. Playing to their strengths is always the way to go.

The following attributes are desirable for a facilitator to harness rich insights for the business:

- demonstrates a belief in the Learning Team process and its possibilities
- challenges the status quo regarding the current state of knowledge and collective knowledge (Work as Done vs Work as Imagined)
- promotes a fair, open and inclusive environment where people feel safe to contribute and explore differing ideas
- remains neutral on the content and does not take sides or express or advocate for a particular point of view
- ensures everyone has the opportunity to contribute equally
- builds trust and respect between members of the group to encourage conversation and learning
- encourages outlier ideas and opinions
- cultivates cultural awareness within the group

- identifies critical connections and patterns in information
- integrates diverse themes and lines of reasoning to create new insights
- understands Learning Team principles and how they are applied.

Choose facilitators for their skill rather than their position. Train people within the work teams to run their own learning sessions. Health and safety reps are an excellent option to train as facilitators.

The skills required to run an effective Learning Team

One client decided to complete a review with their current Learning Team facilitators before training a new batch. The facilitators discussed the amount of time required, from planning to delivering the report.

They also shared insights into what makes the Learning Team a success. The make-up of the participants was vital. Good planning and support enable positive learning outcomes. Developing a safe, trusting environment is fundamental to this role.

Make sure you allow facilitators the time necessary to do a good job. Planning the session, liaising with the sponsor, running the sessions and writing a good report with follow-up is quite a task. Learning Team exercises are pretty tiring, so avoid running back-to-back sessions in a day. Where you're working with several groups, book successive days to keep your energy up.

The purpose is to lead people through a process towards agreed objectives by encouraging participation, ownership and reflection by all involved. A good facilitator possesses the following skills:

- well-prepared
- clear communication
- active listener
- asks questions
- effective at timekeeping
- establishes a psychologically safe environment for sharing
- creates focus in the group
- recognises and manages their biases
- manages the group decision-making process.

A Learning Team lives and dies off good preparation. It's not just about good skills during the sessions; it's about constructing a space for self-determined change. This requires a bit of forethought and the ability to pull together the right people into the right positions.

From start to finish, focus on how to end with every party thinking the process was worthwhile and being keen to do it all again. This defines success rather than a perfect outcome. Everyone feels heard, including those sponsoring or mandating the Learning Team. Everyone's work lives are improved across every level of the organisation.

During the sessions, it's not your job to look good or show how fantastic you are. It's your job to bring everyone together to demonstrate how good *they* are and how fantastic *they* are at what they do. The art of good facilitation is drawing out the great things the team are doing to adapt to their conditions.

Extract their recommendations for what would support them to be more successful and resilient. When the bosses hear these things, they are usually stoked to learn how they can be better managers and gain work improvements they hadn't recognised. This is the win-win we've been talking about.

In practice, this is truly listening – and not just to the words. Understand what teams are saying and draw inferences about their reality. Put aside your biases, ideas and personal leanings. Remaining neutral can be surprisingly hard, but you are simply facilitating.

Coaching others to facilitate is only possible by sitting with them and facilitating together. I often notice first-time facilitators trying to provide answers, giving suggestions, or sharing their viewpoints when the teams are quiet over a question. The trick is to master the art of sitting in uncomfortable silence.

When teams go quiet, several possible things may be happening. They may not quite understand the question, or perhaps they are mulling over whether to be honest about their answer. Or there may simply be no answer. Rather than jumping into the silence with your opinions on how you think they should see things or trying to lead them, you're best to discover the reason by reframing or moving on.

That doesn't mean you only ever ask questions. You can help by distilling the conversation and checking your understanding of the meaning and intent. Encourage them, including those who are shy about speaking up. Join in with the humour and show empathy for their struggles.

It takes practice to be a great facilitator who lets the team lead the conversation. Create a space they can co-opt to use to 'sense and respond'. They can share what they 'sense' and 'respond' with solutions that will work for them. Your job is to trust that the insights will emerge, which means letting go of control of the session.

As you move towards self-directed work teams, the Learning Teams will, over time, have given them practice in operating in an environment of trust with the ability to own their own safety.

The all-important ingredient: trust

Gaining and building trust is the foundation of a successful learning session. Participants need to trust the process – and they need to trust you. Understanding how they deal with conditions means they'll tell you stuff that might have broken the rules. They need to feel safe to share what's really happening – despite the politics of the situation.

To create trust, choose the right sponsor. The sponsor mandates the process and can ensure things get done. They must be supportive and keen to learn from the team. In situations where the obvious sponsor (from a role or positional point of view) is unlikely to be supportive, use a few tactics to find a sponsor who will.

Ensure the sponsor attends the start of the first learning meeting and explains how keen they are to listen and encourage everyone to speak up and be honest, warts and all. This will go a long way to building trust. So, too, will the assurance that there will be no adverse consequences for anything shared. Keep this promise to ensure trust in future and ongoing learning meetings.

Part of establishing trust with the team is having some self-awareness about how they will see you and adapting accordingly. As a white, middle-class, fifty-year-old, teams of younger workers will see me in a particular way. It's common lore in the Safety II world that women are often the best facilitators – ha!

However, I employ a few techniques to break the ice that work for me. I always wear casual clothes, even a hoodie, rather than the trendy, upmarket people and culture look adopted by many senior safety professionals. Find what works for you.

Most work groups have people from a variety of ethnicities, so I introduce myself in Te Reo Māori, the language of the indigenous people of New Zealand. I know enough to introduce myself with several sentences and I went to night school for three months to ensure I could pronounce the words half decently. I can also say hello in most of the Pacific and Asian languages of the countries that surround New Zealand.

> Having facilitated loads of Learning Teams, I find they are always a great experience. As a facilitator, you learn something new every time. Not every Learning Team goes perfectly, but you learn to be pretty good at it over time. Practice makes perfect.

I always finish a learning process with a circle of trust. (I never call it that; it's just a final debrief meeting for the groups I work with, however it is designed to build trust for the future.) Rather than being the middleman, I wrap up by bringing the work teams and their management together, where I present the findings verbally and discuss them in more depth. This means there's no misinterpretation, and managers hear the output with no pandering.

When the workers sit with you and hear what you're saying back to the manager, there are usually smiles all around. As a facilitator, it can be a touching and fulfilling moment at the end of these meetings to see that you've made a real difference for the team. It's worth the effort involved.

In one Learning Team I facilitated, there had been a real breakdown in trust between managers and workers. The participants were reluctant to trust me, thinking I was perhaps a spy sent to uncover details that could be used against them. The circle of trust meeting at the end cleared all this up. It started a new conversation, and a year later, the same workers told me they valued their employer as one of the best in the industry.

These days, I never run a Learning Team without this final meeting. You get far better results if everyone hears it together. Relying on a report to convey the information means ownership of the change is watered down. What's required is a quick, efficient, honest conversation followed by the report to record the discussion.

One word of caution. Never ever have the team's manager or supervisor in the meetings. It can be tempting because they are in the know and are usually helpful people, but you won't get the best intel from the team. This is the *workers'* session. Some supervisors get nervous, so assure them you won't tolerate any personal comments about them. It's simply about the work and its optimisation. HR issues are strictly left at the door.

To exemplify this, once, early in my days of running Learning Teams, one manager insisted on joining, saying he had a great, open relationship with the team and it would be no problem. The problem is, all other things being equal, teams are used to deferring to managers when it comes to problem-solving, and since it's the usual dynamic, they'll do the same in the sessions, more or less.

This manager wouldn't take no for an answer and convinced my client he should be there out of genuine enthusiasm for the topic and methodology. During the session, half of the comments related to what a fantastic boss he was, how he was great at solving problems, and that the team were all happy with everything. The sessions were a waste of time with nothing new learnt. Case in point.

Once the facilitator is confident in their role and understands how to build trust, the next stage is to become proficient with the Learning Team process.

How to facilitate a Learning Team

Conducting a Learning Team is a fairly straightforward process if you follow the general order of steps espoused originally by Todd Conklin. There may be twists and turns that differ slightly, but if you cover these touchpoints, you'll get to your destination – newfound trust between your teams and their managers.

LEARNING TEAM PROCESS

9. FOLLOW UP
The sponsor is responsible for guiding and supporting improvements

8. COMMUNICATION
Ensure the outcomes reach the appropriate people

6. DRAFT REPORT
Analyse the capacities, conditions and constraints

7. DEBRIEF
Make sure you reflect afterwards with the sponsor

5. CONDUCT LEARNING TEAM
Two short and simple sessions (see your Facilitator Guide for how to run these)

4. SELECT PARTICIPANTS
Choose 5 to 8 team members (include a scribe)

2. ENGAGE LEADERSHIP
Find a sponsor and get them on board

3. SELECT FACILITATOR
Choose a neutral facilitator to lead the process

1. TOPIC
A clear topic and scope is the place to start

Figure 17: Your Learning Team process

Preparing and running the Learning Team

In the last chapter, we talked about establishing the need for the Learning Team. A clear scope or topic is essential for keeping on track and ensuring a good outcome.

Usually, you will choose an area of work or a specific task catalysed by an incident or series of incidents. Reframe the topic of the Learning Team according to the nature of the work to be explored. Keep away from discussing safety if possible, that will arise organically when the work is discussed.

For example, if a group of electrical apprentices received more electric shocks than is reasonable, we wouldn't ask what is causing the shocks. Instead, we'd discuss the nature of their daily work. How is the work arranged? How are they trained and supervised? What makes life on-site easy or hard?

We would find out how the apprentices adapt to the challenging construction environment and how they conduct the work. The causes of shocks will emerge out of that conversation. That is the art of good Safety II facilitation – and it requires practice.

As it turned out, that was the topic of one Learning Team. There were challenges with a difference in how the electrical apprentices were taught to carry out isolation processes and the reality when on-site working with their supervisor. They reported that they had to go along with what the person running the job wanted rather than meet the expectations set out in their apprenticeship training. The pressure to do this was felt deeply with a generally held belief that they could lose their apprenticeship placements if they pushed back. Clearly, some had had this experience and it was a common fear.

They navigated the situation by developing close relationships with the host companies they worked with. They supported their commercial interests and needs while learning, as young people, to establish their place on the team and ask for the support they needed. Traditionally, apprentice tradespeople have to endure being put down and ignored on sites, so the challenge was to go against this culture. The Learning Team revealed support from the apprenticeship training organisation that was working and improvements that could be made in future.

The Learning Teams discussion is about how the work is going. As your journey matures, you will get away from incidents as the trigger point for the sessions. Instead, it will be driven by the team wanting to improve challenging work and using the methodology to get the desired results. One operations team asks their safety lead to facilitate sessions when projects are not going according to specification, budget or timeframe. That means weak signals are picked up and solutions are created before something goes wrong.

Generally, two two-hour sessions are enough time for the discussion to unfold. Many do it in half the time once everyone is used to how the sessions run. The first session focuses on appreciative enquiry and the second on developing solutions. Splitting the sessions gives participants time to reflect and have further conversations. Todd Conklin calls this 'soak time'. It's only fair to give them time to review their recommendations.

Getting teams to put forward solutions is about encouraging them to think about what will enable future performance. I often ask them to imagine having an informal, confidential beer with the CEO. If they could say whatever they liked, what would they recommend? I also ask: if they had access to, say, $50,000, what would they spend it on at work? That gets their imaginations going!

Sessions can easily be held via Zoom, as long as each person has their own Zoom frame and sits close-up with their video camera on. Otherwise, make it close to the work environment where work teams feel comfortable. Have a co-facilitator or scribe or use technology such as Miro to record the information quickly and efficiently. It's important to get all the comments down quickly during the sessions in the participants' language.

If you choose to use AI to support your workshop and help produce the report, I recommend a mixture of Miro Assist and Microsoft Copilot to analyse the teams' insights.

Analysing the intelligence

The success of a good Learning Team lies in drawing out the themes effectively. After the first session, create a table with two columns entitled *Helps* and *Hinders*, perhaps on a Miro board. Note relevant comments and insights from the team under each column. From here, draw out the top five or more emerging themes. Map these to the capacities and demands mentioned in Chapter Two. Refer to the horticulture example in the last chapter to see the five conditions identified.

If using AI, run a first draft of the analysis on the spot and review it with the team. I use the keywords 'How are the team adapting to conditions successfully?'. It usually gets a pretty good result. You'll still have to apply your critical thinking and understanding of Safety II principles to shape the AI content into a reflection of what the team is saying. Use your empathy and knowledge to do this.

A primary theme will emerge in most Learning Teams alongside several important secondary themes. When Nippin Anand did a well-received NZISM roadshow around New Zealand in 2023, he talked about Rob Long's 'social psychology of risk'.[9] In the electrical apprentices' example, this turned out to be the major finding. Apprentices felt they couldn't speak up for fear of being labelled unhelpful, sent back to the training provider and replaced by another apprentice. This drove a culture where apprentices (who need patient support and guidance to learn good habits on the job) felt pressured to take the shortcuts encouraged by their superiors. This social aspect of work is often prevalent in the findings of Learning Teams.

The following themes (conditions) emerged regarding how the apprentices create success despite the industry challenges. Table 5 shows how they map against capacities and demands.

Conditions/Constraints	Capacities/ Demands
Support from the apprenticeship organisation is vital to successful work. Apprentice training and education are good but need an adequate feedback loop.	Knowledge Supervision
Apprentices do their best to follow industry standards, although actual practice and industry standards can differ from testing and isolation procedures.	Work method
Political dynamics and commercial realities create challenges with speaking up. They grow in confidence when they challenge the status quo on sites.	Culture Goal conflicts
It's vital to have good quality tools that are fit for the job.	Tools and equipment
Psychosocial risks are inherent in construction culture. Apprentices try to report these. They also rely on informal peer support to resolve difficult situations.	Culture

Table 5: Mapping conditions and constraints to capacities and demands

Once these conditions and constraints have been identified, the final stage is to identify the top three challenges and the recommendations put forward by the teams. This prioritisation should be done in collaboration with the team. Using the Miro Assist function, I use the keywords 'challenges and solutions' to analyse the data and start to build the report.

In this case, the teams had some great suggestions for the apprenticeship organisation. These centred around:

- supporting a 'speaking up' culture on site
- collaborating with the industry regarding the discrepancy in standards for isolation procedures
- redesigning apprentice training to reduce information overload and increase retention.

The apprentices detailed how each of these could be achieved. They presented to their sponsors on each point so that further questions could be asked and matters clarified. The great news is that, a year later, electrical shocks are down by 50%!

A quick final note about framing for success and how to do it: occasionally, you'll encounter a Learning Team where people feel pretty disempowered. They'll whinge and moan and generally express a bit of negativity. What's happening here is that they aren't confident in their ability to 'sense and enable an appropriate response'. Fair enough.

So, we need to listen to those frustrations and use critical thinking to understand where the success adaption lies and how this should be further supported. In the case of the apprentices, some were handling the challenges onsite well, and others not so well (construction sites are still pretty tough working environments). It was a matter of looking at what those who were successful had learnt and finding ways to support others to reach this success too. Also, make sure to ask what the bosses are doing that helps the team, so positive feedback can be given.

Setting up the organisation for success

The final challenge is to deliver the messages back to the organisation so that everyone wants to continue learning together.

The first task is developing a report to highlight the findings (see Figure 18). A longer form report will be needed for more weighty Learning Teams to ensure the necessary detail is provided to leadership. Nowadays, I tend to build the report on a Miro board using the built-in functions. The output can be exported as a PDF to send to colleagues.

To maintain confidentiality and build trust in the process, run the draft report past the participants in the group to check your understanding and clarify any points. As mentioned earlier, a face-to-face 'circle of trust' is the best way to report the findings first.

Finally, the organisation must follow up and take action on the findings. It is the ultimate evidence of listening and means participants in future learning sessions will continue contributing effectively.

Findings often warrant short follow-up learning sessions to gather more detail and dive further into the solutions needed.

LEARNING TEAM SUMMARY AND OUTCOMES

REASON FOR LEARNING TEAM:	DATE:	FACILITATOR: SPONSOR:	MEMBERS:

BACKGROUND TO CURRENT SITUATION

	DETAILS:

KEY POINTS FROM DISCUSSION
MEETING ONE

HELPS:	HINDERS:

KEY THEMES
CONDITIONS AND CONSTRAINTS

CHALLENGES IDENTIFIED AND RECOMMENDED SOLUTIONS
MEETING TWO

CHALLENGE	SOLUTIONS	RESPONSIBLE

Figure 18: Your Learning Team process

Establishing cultural safety as a facilitator

In a Learning Team with a construction company that employs many Filipino carpenters, we asked a Filipino quantity surveyor to join the group to translate and support an open conversation. He assured the team how the information would be used. They trusted that he knew the background conversations in the lead-up to the group session. Great intelligence was shared which the management team were very grateful to hear about.

When you have a culturally diverse group and many whose English is limited, involve a trained interpreter from the dominant languages of the workers. The questions can be translated into other languages, but having a trusted support person who understands the nuances of language and culture is essential.

If I struggle to find such a person, my backup is recordings from workers from diverse languages explaining what a Learning Team is and the desired outcome. They explain that no one will be fired for speaking up and that it's a great experience. In one of my videos, the person said they initially thought the Learning Team could be a set-up but confirmed it was not.

I've found that success framing works well across cultures. When they get a bit shy, I ask them to tell me what their bosses are doing well. That usually gets them smiling and talking. After they share a positive comment, I ask, 'And how could we improve this?' Again, they are forthcoming. I keep repeating this question until the team are comfortable sharing any difficulties, knowing they've already been positive toward their bosses.

Providing appropriate food sets a welcoming scene, as does starting and ending the session with an appropriate prayer. Learning the group's cultural norms in advance helps you connect in a way that is comfortable for them. Using greetings in their languages makes them feel you care.

All facilitators should attend cultural competency training. The crux is to get alongside the group. Treating the team with respect and kindness goes a long way to establishing trust. These values are universal, and your efforts to connect will be appreciated. The platinum rule for connecting across cultures is to treat people as they want to be treated.

When facilitating the horticulture sessions, we involved government liaison officers who provided translation and supported cultural safety to ensure the seasonal teams from the Pacific Islands felt okay to speak up.

In my experience, the key to successful Learning Teams with diverse cultures is how you handle the sponsor and the operational management involved. As discussed in Chapter Six, cultural competence training supports their ability to listen well across cultures and develop the empathy to collaborate with work teams who often defer to authority without question.

I hope running proactive Learning Teams means you will never have incidents again. However, if you do, the next chapter covers how to support a team when something goes wrong.

CHAPTER 9

Learning Reviews

Shift from blame to a learning culture

Key points

- The New View of human error provides alternate ways to view incidents and events.
- Local rationality tells us that all human action is governed by what made sense at the time.
- Complexity theory says a holistic view of events provides better organisational learning.

A couple of years ago, I received an email from a colleague whose company had just had an incident. She wanted to know when she should issue a warning to the worker involved – before or after the Learning Review. She explained that all their documented procedures were in place, but the worker hadn't bothered to follow one. (Note the loaded language.)

Admittedly, the Safety Differently ideas were new then, so this was a genuine question. Clearly, though, she didn't fully understand the purpose of a Learning Review. She had already decided who was at fault.

She went on to explain that the incident hadn't come to light until senior managers visited the site with clients some days later. Unfortunately, the political context around an incident influences how it's responded to.

My colleague had a well-intentioned desire to do the right thing and find a way to prevent incidents. However, using blame and a stick to improve the situation would likely result in further serious incidents occurring and not being reported. That had probably been happening for some time, and issues were not being adequately resolved. This kind of vicious spiral gets us nowhere. Many of you will likely have similar stories. It's easy to jump to blame the work team who, at first glance, seem to be at fault.

There are plenty of good books written on this topic. To get the best, read Sidney Dekker's teachings, or – even better – take one of his workshops or university papers. It's beyond the scope of this book to delve deeply into better responses to accidents. However, it's an important part of the mix because we're focusing on enabling team resilience.

Remember that if you're only doing reactive Learning Teams, you'll need to shift to a proactive focus to pre-empt things going wrong.

Human error and the politics of blame

We live in a world strewn with terrible and tragic disasters. It's one of life's enigmas to understand why they happen. As a society, we've wrestled with the question: are all accidents preventable? Notwithstanding a discussion on the esoteric nature of this question, the social harm and cost of accidents warrant every attempt to prevent them.

> We live in a world with real accountability issues, but hanging people out to dry never works. Every team member feels judged and mistrusted when written warnings are issued, and silence can ensue.

In the example above, the first task was to build trust and share information about what was truly happening at that site. How was everyday work going? What conditions and constraints were they regularly facing? How were they creating success each day? The insights from these questions would create a new two-way conversation where problems would be revealed and solved. Despite management's best intentions, they would inadvertently worsen the situation by issuing a warning. If there wasn't enough trust to sort out a serious issue, what else was affected in the organisation?

To look at where we want to head, let's quickly consider the central theme from Sidney Dekker's seminal book, *A Field Guide to Understanding Human Error*, in which he compares old and new views of human error to understand what we need to learn when an event occurs.[10]

Dekker tells us that human error is a *symptom* of deeper trouble, rather than the *cause* of trouble. It's easy to get caught up in the need for accountability; however, it's more often the organisation that set the person up for failure. We must set aside our hindsight bias and fathom what made sense to the team when the event occurred.

Sensemaking and local rationality

The discovery part of the Learning Team is about understanding what *actually* happened — not what *could* or *should* have happened. Not *why*, but what made sense at the time.

THE TUNNEL

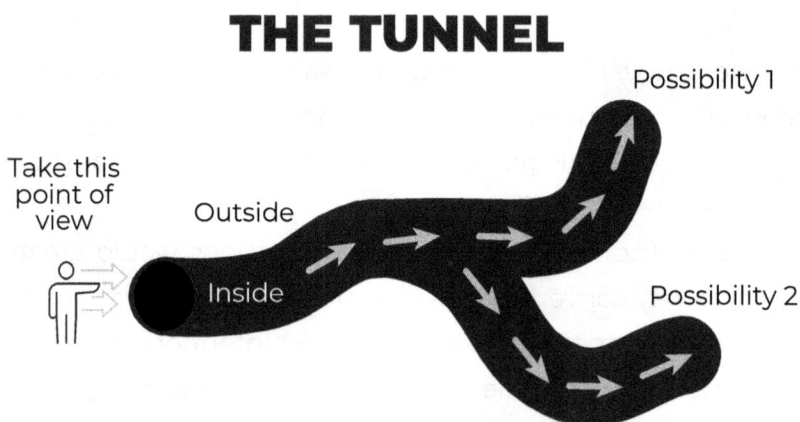

Figure 19: Making sense of possibilities – Sidney Dekker

We call this local rationality (what was rational to the person local to the work), and the best way to understand it is by getting into Sidney Dekker's tunnel.[11]

The tunnel (Figure 19) is a metaphor for seeing what the worker was thinking as the job unfolded. They could only understand events based on what they knew and could see at the time. No one has a perfect data set of what's happening around them as they go about their day. We don't have eyes in the backs of our heads.

In his regular talks, HOP coach Bob Edwards explains: 'We need to learn enough to realise that, given the conditions people face, the information they hold, the tools and equipment they have and the pressures they're under, we would probably have made the same decisions'. This helps us to understand the complexity of the situation.

A well-known example is the movie rendition of how Captain 'Sully' Sullenberger crash-landed a US Airways Airbus A320 on the Hudson River in New York in 2010. The movie showed how the investigators painted a picture of events that suited the political needs of the situation. The reality was that, given the brief timeframe to make decisions, Sully made exceptional decisions under pressure and most pilots could not have achieved the same outcome.

Another example in recent years was when a driver was fired for accidentally running his double-decker bus into a single-story garage. The incident made front-page news in New Zealand. I posted it on my LinkedIn page, asking for comment and got an international flurry of comments supporting the bus driver.

The New Zealand Employment Relations Authority upheld the dismissal despite, as his union pointed out, five other bus drivers who did precisely the same thing yet managed to keep their jobs. I sent the outpouring of LinkedIn messages to the union. A chat with the union secretary followed, and lawyers were called to raise an appeal on the matter.

The challenge here is that even the justice system, our regulators, and many other systems that support social accountability are sometimes limited in ensuring an appropriate response. Neuroscience shows that some mistakes are linked to autopilot (like the bus driver situation) or other cognitive brain functions. Blame and punishment are entirely inappropriate.

Just look up the material on 'forgotten baby syndrome'. In the worst of all tragedies, autopilot causes a parent to leave a sleeping baby or young child locked in a car, forgetting they are there. Should this parent face criminal charges for negligence in this situation?

Workers read cues that their senses pick up – all day, every day. Invariably, mistakes beyond their control will occur. In addition, error traps exist to exacerbate this, such as being overworked, stressed, in conflict with others, or confusing instructions and dials, which puts more cognitive load on people than usual.

Performance modes demonstrate that the human brain operates efficiently based on the level of cognitive demand required. That is the reason for what we call complacency. Over time, our brain seeks out situations where it judges less cognitive processing is required. Given the vast amount of data we process every day, it is a normal human function. We can't blame people for their human physiology.

So, now that we're keen to learn instead of blame, how do we go about it?

Find new language

The first step in shifting organisations from leaping to blame is to rethink your language. While it may not seem like a big deal, language does a lot to frame ideas in our minds. Safety I terms such as 'investigation', 'incident', 'root cause' and 'corrective actions' have become so commonplace in recent decades, that they evoke assumptions that errors are bad. That leads to blame.

The first and most important language change is from the rather strict and formal sounding 'investigation'. It is now common to use 'Learning Review', which has a far more inclusive connotation.

Some of the new language can sound a bit flowery and may not be accepted as well as New View practitioners might like, so feel free to come up with your own terms or choose what works with your people. Other options could include 'learning conversation', 'event', 'condition', 'sensemaking' and 'learnings'. Be creative.

Use a Learning Review for near misses or less serious incidents

Try to hold the Learning Team as soon as practical after the incident, while information and memory are still fresh. Always start by checking in on how the people involved are feeling – some will be in the room while others will not. Allow people to tell their own stories about what happened. Everyone's stories matter.

The fact is that in these situations, people will be feeling terrible about what happened. Make sure you keep the tough part of the conversation brief and quick. Spending too much time pressing against the wounds will only make everyone feel bad.

Some great questions to ask include:

- What could have happened?
- What worked well?
- What surprised us?
- How did our system set a worker up for this failure?

As a group, work out what conditions were present when the incident occurred. For example, I often see incidents in industries where fatigue is a factor, e.g., the transport sector. Fatigue, of course, can be intangible and Safely I style fatigue management protocols only deal with part of the problem.

The conditions emerge out of an understanding of the work dynamics. How was the work planned, organised and executed? How do we handle a difficult day? What do you need to make the day go well? Ask how people were feeling ('I was tired') or about the circumstances ('I pushed the wrong button'). In our example, understand how travel routes are determined, how scheduling and client interface occurs and what environmental constraints people face.

The actions of some people in the room may have contributed to the event, but remember, most actions won't have been deliberate or malicious – it's often a lapse or mistake. A Learning Team probably isn't the right approach if the behaviour was reckless. Prioritise those areas that require improvement and the actions to address these.

You can still do a timeline, but try to do it chronologically to help you stay in Sidney Dekker's tunnel and see the trajectory of events from the workers' perspective.

HOW DOES WORK LIKE THIS NORMALLY HAPPEN?

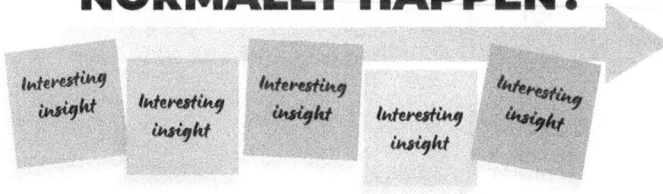

HOW DID WORK HAPPEN ON THE DAY OF THE INCIDENT?

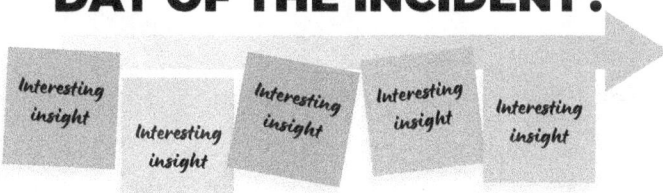

Figure 20: Building your Learning Review timeline

The next stage of the analysis is to understand the conditions present throughout the trajectory of events.

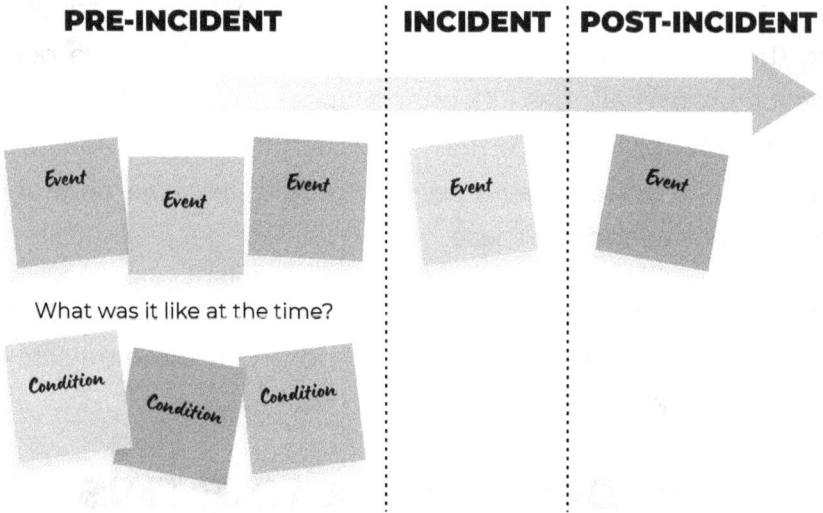

PRE-INCIDENT | **INCIDENT** | **POST-INCIDENT**

Event

Event

Event

Event

Event

What was it like at the time?

Condition

Condition

Condition

Figure 21: Understanding conditions present which triggered the event

If a stakeholder still requires a report using root cause analysis, such as ICAM (Incident Cause Analysis Method), it's still a relevant way of analysing an incident. After the discovery process and during the analysis stage, map the conditions to a PEEPO chart or the four ICAM dimensions for further understanding and review.

Sometimes, a traditional incident investigation may have gone ahead without conducting a Learning Team. A good option is to follow on from the investigation with a proactive 'learning from normal work' Learning Team based on the general work area involved. This follows the process of the last two chapters.

Conducting Learning Reviews for serious events

Occasionally, I get calls from safety leads who've had an accident where authorities are likely to be involved. They want to know if a Learning Team is appropriate and useful in this situation. I've had this occur for a serious national event that needed to be handled very sensitively.

In a severe incident, consider the legal risks and implications for stakeholders. There are various options for using a Learning Review in this situation. Depending on your circumstances, you will have to make the judgement call under pressure and collaborate with your leadership team.

In the national scenario just mentioned, I recommended involving kaumātua instead of using an in-house facilitator. In Māori culture, this is a senior respected member of the community who can act as a neutral party and handle the sensitivity of the situation with maturity. When emotions run high, the company will naturally worry about damage control, and workers will have fears and anxieties that are difficult to voice in a highly charged situation.

Another client called me when a serious crushing incident had occurred on site. Their client knew they did Learning Teams regularly and immediately demanded they do one and include the client in the session.

To reiterate: this is *not* how Learning Teams should be used, as the risk of the session being used to apportion blame is too great. The aim is to construct a psychologically safe environment where everyone can share.

So, how do we use a Learning Team at these times when so much is at stake? There are a couple of options. The first involves a traditional investigation and quietly running a Learning Team to find out what happened and deal with any issues under the radar. This could even be conducted verbally without documenting the session. The second option involves the same principle, but running the Learning Team openly while using legal privilege to support a clear and honest conversation.

The need for justice – handling the difficulties of accountability

When the proverbial hits the fan, organisations tend to panic and react. Anger, fear and grief quickly override a measured approach to learning.

> A Learning Review approach involves holding out the jury. It's hard to do in serious situations. Grappling with the need for accountability versus the need to ensure an open, transparent and just culture presents challenges when people want answers.

For the most part, it's what Sidney Dekker calls forward-looking accountability, where everyone who is actively involved becomes part of the accountability process.[12] The outcomes focus on setting everyone up for success by understanding what future capacities are required.

We also need to consider restorative justice when someone has been seriously hurt or killed. It's not only the safety industry that is adopting these principles. In simple terms, this is where all parties get together and get answers and emotional closure in an atmosphere of openness and vulnerability. People are heard when strong emotions are involved.

I've witnessed a restorative justice process in a fatality when the company was definitely negligent. Lessons were learned instead of hidden and ignored. The company practised humility and the family received the respect and kindness they deserved over a number of years – which was perhaps the most important aspect of the process.

Let's turn now to an added dynamic with our learning approaches. How do we include contractors?

CHAPTER 10

Contractor Partnerships

Co-operation and trust lead to business value

Key points

- A partnership, rather than a top-down relationship, pays dividends with your contractors. Build trust and transparency into the way you work together.
- Bringing your contractors into your New View safety culture is crucial to success.
- Collaborate with contractors by running your Learning Teams with them.

A client who wanted to get going on Safety II was a major contractor in a large rail infrastructure project. Their leadership had one of the best appetites for change I'd seen, so I was excited. First, we asked the whole team if they wanted to do Safety II. They said yes. Their client, a well-respected organisation with access to a massive budget, had also been invited to the conversation.

As a follow-up, the client called a meeting with the contractor, saying they wanted to support the process proactively and asking what more they could do. This is the best-case scenario we're aiming for. The company went on to do very well with New View approaches and became a finalist for the 2024 national Safety II award in New Zealand.

Another construction client wanted to increase trust and transparency with their main contractor. They paid for the contractor to run a Learning Team and find out what was helping and hindering performance on the site. After some constructive sessions, it emerged that how the client managed the job and handled variations was affecting the work teams.

Despite support from their client, the contractor was nervous about openly discussing how the variations affected them. They were happy to be open about their own failings in the spirit of the exercise, but when it came to pointing out the failings of their client, they went quiet. When someone is paying you, it's natural to fear the ramifications of honesty. In this case, it meant the directors at the top of the food chain, didn't get to see how their decision-making was flowing down the various layers to the crews actually doing the building work. Really, what was needed was the opportunity to listen and change how they were doing things for the benefit of all.

The challenges and rewards of doing New View Safety with contractors

From a commercial perspective, the contracting relationship is designed to provide expert resource on an 'all care, no responsibility' basis. The arm's length nature of the arrangement tends to cause trouble with the inherent need for building trust that is imperative to the New View approach.

One organisation that adopted the Safety Differently ethos became highly collaborative and engaged with safety, meaning monitoring was no longer necessary (yes, it's doable). However, whenever a contractor came to work with the team, their attitudes to safety gleaned elsewhere were part of the package, sometimes resulting in unacceptable safety standards. This situation frustrated everyone.

There is no other way to cut through this. To reap the benefits of better trust and collaboration, you must work with your contractors via a partnership approach. One way to do so is to involve them in Learning Teams, which build trust and create a shared problem-solving environment. It is the difference between *being* included and *feeling* included.

Another organisation I worked with wanted to conduct their first Learning Team on a troubling repeat event. A particular site was chosen because the senior operations manager would likely support the Learning Team approach, but most of the work was conducted by contractors.

The workers – a fine bunch of young men who operate excavators for a living – were asked to join the Learning Team. After multiple assurances that the truth behind the incidents would be responded to respectfully, they were smart and forthcoming with ideas for improvement.

We discovered that, due to uncertain market conditions, the company had tightened its belt following COVID restrictions. This action had a trickle-down effect on resourcing the teams with vital technology and processes. The power distance and divide between management and contractors meant the intel wasn't getting back to senior leadership in this company. The contractors, business owners in their own right, had great suggestions for resolving the issues.

Bringing contractors into your Learning Teams

Your Learning Team programme is about understanding how to support safe and successful work. That, of course, must include the contractors who do your work. The question is how to do so.

> Organisations sometimes baulk at the cost of bringing contractors into Learning Teams. However, they contribute insights and ideas that are worth the effort.

Their perspective will likely be different from your employees. Because they are self-employed, I generally find contractors activate their business acumen on your behalf. In addition, they see how other organisations in your industry are doing things and can bring good ideas across.

There's a caveat, though. To build trust and support a mutually beneficial relationship, recognise that contractors will struggle to tell you the straightforward truth even more than your employees. No one wants to bite the hand that feeds them.

The reason for this is obvious – and not to be underestimated if you want good results from the Learning Team and your ongoing relationship. As the adage goes, you're as good as your last job. Contractors risk losing income if they tell you the truth.

That means you must work even harder to create psychological safety and assure the Learning Team that you are prepared to be vulnerable. If any unwelcome truths arise, resorting to your usual forms of power over your contractors will undo all the good work.

> To be truly forward thinking, introduce the learning processes in your contracts. It's a great way to set up a no-blame approach.

A large budget facilities and maintenance team for one of New Zealand's governmental organisations wanted to trial Learning Teams; however, most of their work was carried out by contractors. To keep it simple, we decided to use the requirement for an annual contractor review to hold a Learning

Team alongside the traditional top-down performance style review.

A selection of primary contractors was bought together (at their own cost). They were asked how they handled capacities and demands when working within the facilities and its services. Many insightful learnings emerged about how they were working around the systems and processes to get the job done.

It also became clear that there was confusion around protocols relating to several critical risks. The facilities team recognised that traditional contractor reviews tended to be one-way conversations, so instead, the Learning Team approach uncovered pertinent issues that needed to be resolved. It was a good exercise for both parties and led to better collaboration.

The head of facilities and safety manager shared that this approach built engagement with safety from the contracting teams, who were tired of over-the-top rules and onerous paperwork from their many clients. As a result, the relationship improved and, even better, the safety insurance auditors wrote a rave review on the approach.

Handling Safety II for lower tier contractors

I've worked with numerous contractors who work for principals in the contracting food chain (e.g., councils and government departments). These contractors can be sizeable companies in their own right and are often highly capable.

These companies get very annoyed at the extent of the monitoring undertaken. They are hired for their expertise and are then monitored by third parties with less capability. It becomes a battle of 'managing upwards' while getting the job done within spec, deadlines and budgets.

Some years back, a friend with a strong professional reputation had a team of about fifteen electricians working with him in his business. He and his wife worked all day in the field, only to come home to hours of paperwork every night. Increasing demands for administration kept coming despite the company having no accidents in almost two decades. That admin added no value. Instead, it added stress and many late nights, contributing to fatigue at work. It seriously impacted their quality of life, which, ironically, is contrary to the reason we say we ask for the paperwork.

Let's look at this through a capacity lens. A civil contractor I worked with started running Learning Teams and found that the client (a government department) often came and monitored their sites. The client got angry if particular safety protocols weren't followed and made constant demands of the contractor about workflow and public safety without fully understanding their day-to-day work.

The teams were between a rock and a hard place trying to deliver results while keeping the client happy. Sometimes, it took a fair bit of window dressing, which wasted time and put pressure on the teams. In that sense, the well-meaning client reduced team capacity by withholding curiosity or failing to seek to understand.

It can be challenging for contractors to push back on this. I've seen tears when a progressive contractor was forced to return to traditional methods and modify their wisdom – just to please the client. Having your contractors teach you better ways of doing things might be humbling, but it's a worthwhile experience.

Now that you're bringing the whole wider team into your New View approach, it's time to look at how to build and evolve your learning framework.

CHAPTER 11

Your Agile Learning Team Framework

Build a framework to hear the weak signals in your business

Key points

- A robust framework means, over time, you'll understand what drives capacity and resilience for your teams. This forms the foundation for your due diligence reporting in the next chapter.

- There are a variety of learning approaches that can be adopted.

- Support your learning with freedom in a framework principles.

A Crown Research Institute (CRI) in the New Zealand forestry industry started doing proactive Learning Teams several years ago. Following a training course with the safety representatives in each team, they embraced the approach as a means to collaborate and get better solutions than through top-down decision-making.

At first, Learning Teams happened for everything. What a flurry of activity. Team input into all sorts of things was valued and changes were made. Most of the Learning Teams were initiated by the teams, who also led the changes. 'Freedom within a framework' gives teams the ability to make their own decisions within certain boundaries. It allows work teams to take responsibility for the solutions and builds ownership and trust.

Over time, they settled into an ongoing maintenance level of Learning Teams. You may or may not want a formal framework for this, depending on the size of your organisation and the existing culture. The CRI's approach has contributed to a collaborative culture where teams feel they can influence positive change in their workplace and feel heard.

This ongoing proactive use of Learning Teams is necessary to hear the weak signals in your business and prevent drift to failure over time. Organisations that maintain the approach see fewer incidents and have generally happier staff. Anecdotally, across industries, those who embed Learning Teams as part of their culture report better relationships with teams and operational improvements.

> Done well and consistently, Learning Teams
> can really turn around a business.

We find that organisations that take Learning Teams seriously report similar results. An organic cadence of Learning Teams emerges, driven by the operational teams who request them. The value is reported by all those involved. Usually, the programme is supported by champions in either the safety or HR team. Even leadership can be the driving force. In this case, the safety lead ensures the workshops are maintained and remain visible.

You'll need to design your framework based on your particular critical risks and industry context. For example, one organisation was instrumental in maintaining infrastructure in a large New Zealand city. It needed to ensure strategic objectives were met and wanted to use the Safety Differently methodology to deepen robustness and develop excellence using the latest science and research.

The learning framework needed to combine with industry guidance for risk management. That included how risks were managed across multiple stakeholders with complex interactions between workers, the public and a variety of transportation modes.

Learning Teams will add to your critical risk management processes. You'll need a focus on failing safely with critical controls and a verification programme, including best practice performance standards. Learning Teams will support your programme by developing shared responsibility, consultation across a contracting supply chain, and helping to conduct risk assessments where worker's perspective is vital. This chapter examines elements of a framework for generalised settings.

Building your learning framework

So far, we've covered proactive, post-event and contractor Learning Reviews. However, you can use Learning Teams in many ways, including wellbeing, quality, and change management, even conducting critical control reviews and decluttering a process. Keep in mind the micro-learning methods that robustly track team resilience and are much shorter.

The thing is, you need to focus on getting to a stage where your learning, and therefore collaboration with your teams, is robust. A Learning Team here and there is useful and will be effective at solving a particular problem. But it won't ensure the overall results that safety science prescribes unless you build and implement a framework that is right for your context.

The sample framework in Figure 22 suggests the broader use of learning tools. What guidance might you give your organisation to create the necessary robustness?

SAMPLE FRAMEWORK

Better Work teams	Periodic learning reviews	Proactive learning workshops	Post event learning reviews	Freedom within a framework
· Understand capacities/ demands · Listening to weak signals	· Better Work team reviews · Established cadence	· Critical risks and controls · Success in operations · Difficult work · Other	· Near misses · Incidents · Serious events	· Safe co-design of safety management system

Figure 22: Sample framework of learning tools

One client developed structured guidance to show when to use particular tools to support active use across the organisation.

Figure 23 indicates the potential use and cadence.

LEARNING TEAM GUIDANCE

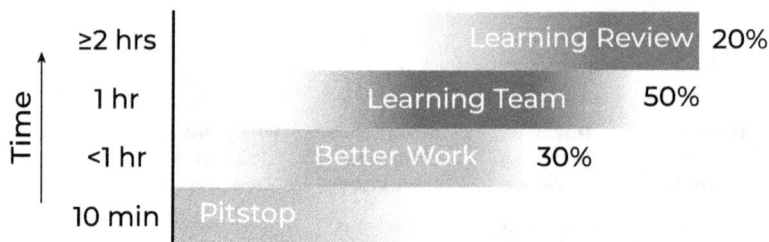

Post event potential severity	Observations & (notifications)	Minor	Moderate	Serious and above
Proactive learning (periodic)	Understanding conditions - weak signals - success - ideas	Understanding work & success - conditions - capacities & demands - weak signals	Understanding work & success - conditions - capacities & demands	
Who facilitates	Anyone, team supervisor	Anyone, team lead	Trained facilitator	Trained facilitator

Figure 23: A possible learning framework

Consider how you'll capture the insights in your current systems. Most people adapt the language and fields on their 'observations' or 'actions' forms to report the insights and items to follow up. The aim here is to keep it simple and not introduce new additional systems – existing ones will do.

Next, we'll tuck into some basic descriptions of the tools required to set up your framework. You will want to innovate with these in your context. The world is your oyster. You are only limited by your imagination.

Everyday learning – Better Work teams

You can develop a tool or set of tools. We have tended to use safety toolboxes to do a lot of *telling*. Why not use them to do more *asking*? Gathering intelligence regularly on what is causing capacity challenges in teams will reveal operational issues that senior leadership should know about. The directors must have transparency over this so they can prioritise and resource improvements accordingly. As we've explored, the frontline sees what's really happening.

We use health and safety committees with team representatives to gather information on safety concerns. However, we gather little insight into what is helping and hindering performance and, therefore, driving safe outcomes.

Safety II tools should be used regularly to capture insights as they emerge. One company holds a three-monthly review session with each site team facilitated by a safety lead. Another has tools on an intranet for workers or for use by a safety lead facilitating a conversation. Regularity is important because Work as Done adapts and morphs quickly. In Australia, these are called 'worker insights'.

I often see organisations stuck with their Safety II approaches because they try to do longer Learning Teams and get bogged down. Innovate your way to finding micro versions so the processes become dynamic. At the start, however, be sure to learn the full version of these approaches so everyone captures the essence and understands what they're trying to achieve. Plus, there will always be times when a topic or issue warrants longer workshop processes.

Some teams spend extraordinary amounts of time on useless risk matrices and paperwork, yet the simplest and smallest tool is the good old-fashioned conversation. (There's even a tool for that.) Training with managers and leaders helps them understand the theory and coaching skills required to conduct this effectively.

Another option is using existing operational processes in conjunction with the resilience principles and models. One client had a process for reviewing completed projects. They looked at the process used and integrated the Safety II principles to understand how success was being created and how to improve team capacity.

The next challenge is filtering all the insights you gather into meaningful data for your leadership and directors to inform their decision-making. That is the topic of the next chapter. In the meantime, we need to cover the more robust Learning Review and other types of Learning Teams.

Learning Reviews to deepen assurance

Learning Teams can be used to review projects and work sites proactively. Utilise this approach alongside traditional audits and inspections to gather collective intelligence from the teams for insights commonly missed by assurance activities. A thorough review would involve work teams, engineers, leading hands and technical experts.

A large-scale New Zealand infrastructure project was struggling with workers regularly breaching procedures. Policing, audits, corrective reporting, and discipline weren't dealing with the issues. The situation clearly indicated that an adaptive challenge was present. The team decided to view these instances of human error as symptoms of deeper problems.

They conducted a project review in collaboration with their clients to understand how the teams were creating success in their work. This involved several levels of the organisation in multiple groups to discover what was helping and hindering team performance. Work teams were involved, plus site management, engineering consultants and project managers from the client side. Many learnings were elicited, and performance improved on the project and with safety.

In these situations, it's important to run the learning sessions according to the level of hierarchy. Participants in workers' sessions should be on the same level. Bosses and other managers should join the session for the next level up. In this situation, the project director would sponsor the Learning Team.

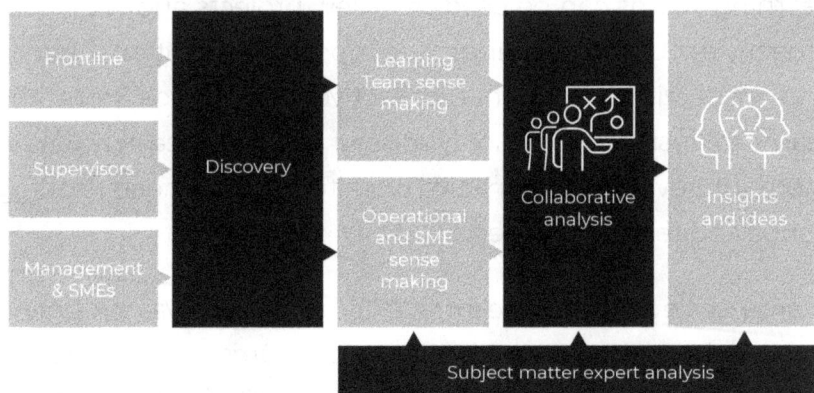

Figure 24: Breadth of learning to undertake in a full review. (Reproduced with permission from Art of Work.)

Another example of this level of review was a client who wanted to take a major near-miss seriously after it was apparent there had been warning signs for a while that somehow were missed. In this instance, they ran a series of learning sessions at differing levels of the organisation's hierarchy to understand how systems and communication channels were working. After sessions with the repairs and maintenance team, middle management and senior leadership, the issues became apparent and a collaborative action plan was created.

Learning Teams to review critical control effectiveness

As we've seen, to support our teams to fail safely, we must ensure critical controls are in place and more importantly, are effective on a daily basis. It requires a new lens to truly understand this and have directors recognise how controls operate in practice. Current methodology tends to emphasise reviews and checks by subject matter experts. It totally misses understanding how the control interacts with work variability and demands.

So, what is a critical control?

> It's a control that is crucial to preventing the event or mitigating the consequences of the event. The absence or failure of a critical control would significantly increase the risk despite the existence of the other controls.[13]

Kelvin Genn, from Art of Work, has developed a fantastic model (Figure 25) to analyse critical controls from a human-centred point of view. With his permission, I share the basic model to show that a New View look at what helps and hinders great work in the field can be a game changer. The following descriptors provide insights into the adaptive capacity of each critical control.

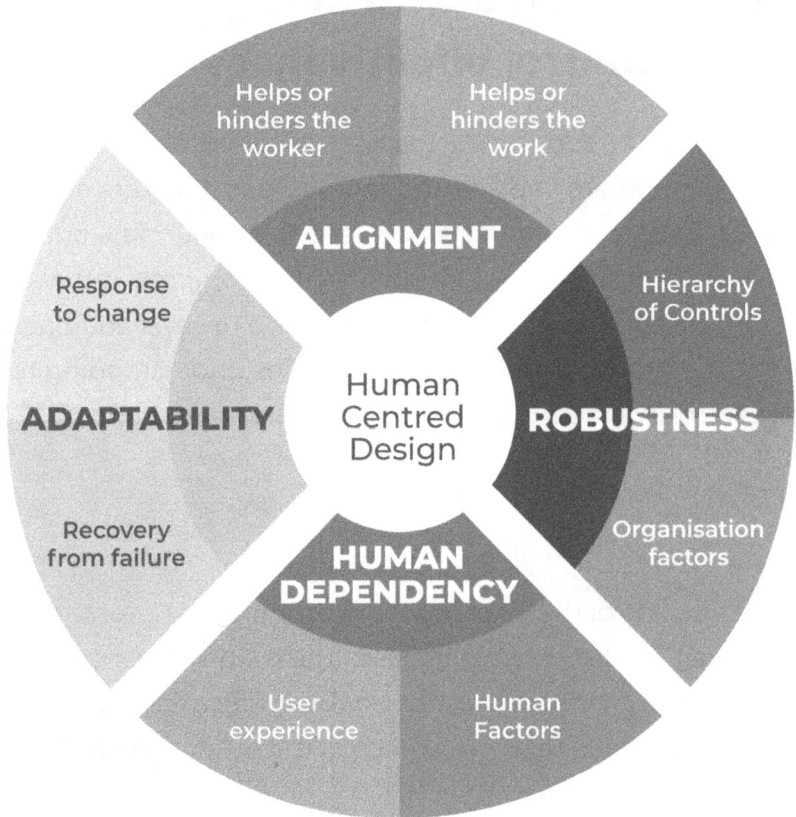

ALIGNMENT	ROBUSTNESS	HUMAN DEPENDENCY	ADAPTABILITY
Whether the control helps or hinders the person who is doing the work and the work itself.	How dependent the control is on the human and human factors that impact the work.	How discoverable, understandable and usable the controls are from the worker's perspective.	How well the control adapts if plans change or unforeseen events take place.

Figure 25: Human-centred design. (Reproduced with permission from Art of Work.)

This model can be used in discussions on critical controls as part of conversations with leaders and Learning Teams, particularly following events, quarterly reviews and deep dives, and adapted into other traditional critical risk management methodologies. The approach adds to your current critical risk programme and can be shaped to your context. We'll look at this in more detail in my next book, *Evolving the New View of Safety*.

For now, let's consider how to use the model as part of your learning framework by building it alongside a Freedom within a Framework approach.

Using Freedom within a Framework

When we describe Safety II as enabling employees to adapt successfully to variability, we know that Freedom within a Framework becomes an essential tool in our toolkit. While the term has been co-opted into New View terminology, it comes from business improvement thinking. Here is an apt definition:

> Freedom within a Framework means providing employees with a context for behaviour and performance. It establishes adaptable parameters that give employees a sense of control and ownership because they know they are not locked into fixed rules.

Here's an example. In the transport industry, drivers often have to climb onto the truck deck to load and unload. It's not ideal, as a fall of even a metre can be nasty. Usually, appropriate tools and equipment are available at a client's premises to help avoid the need to hop up on the back. Sometimes, though, it won't be possible, or adaption is required for other reasons relating to the job. Rather than a hard and fast rule, drivers are expected to exercise caution and only get onto the deck when absolutely necessary. You could write certain protocols into the procedures that must be followed and others where discretion can be exercised.

One client effectively used Freedom within a Framework principles while maintaining the traditional critical control review approach. The client's safety team chose a critical risk each month. Audits and inspections were conducted along with top-down admonitions of what the teams should and shouldn't do. Some of this was useful, while some was a waste of time.

They decided to trial the new critical control review approach with Learning Teams to support the Freedom within a Framework ethos. What emerged was a real eye-opener for management. A whole new set of insights became apparent. The first step was to go back to their client and insist on a more collaborative approach.

My client discovered issues with the critical controls that no amount of inspections and hazard reporting forms could highlight. On top of that, some innovative ways of looking at continuous improvement came out of the woodwork.

A summary of the top three findings and the proposed solutions were as follows:

- **Finding**: Some administrative controls (traffic management plans) are overly prescriptive and rigid, showing poor alignment and adaptability.

Solution: Work constructively with major stakeholders and clients to further understand how these processes reduce the ability to adapt effectively on the job and make changes.

- **Finding**: Availability and resourcing of tools, equipment and consumables affect human dependency and alignment.

  Solution: Collaborate with work teams to investigate having more transparency and authority of spending for team resourcing and necessary work-based expenditure. This would enable real-time adaptability to conditions.

- **Finding**: The traffic management role is demanding and heavily human-dependent. Some conditions reduce alignment.

  Solution: Conduct a Learning Team with these staff to better understand the role from the workers' perspective. Review recruitment, retention and job rotation elements to increase the team's capacity to respond to changing day-to-day conditions.

Learning Teams for change management

One client wanted to make changes to an area with frequent traffic where hazards and risks were created by trucks loading and unloading product, forklift movement, and pedestrians needing to pass through to other parts of the production facility. Needing to handle traffic management as part of the process, they wanted to assess the risks from a technical expert perspective and engage with their teams to gather insights from those working in the area who understood the practical realities.

In this instance, the safety lead had identified the risks and knew the associated regulatory requirements. Team insights, however, showed that some so-called best practice controls worked in practice while others didn't. For example, having a designated place for the truck driver to stand caused other risks in certain circumstances. Being uncompromising about this would introduce other hazards.

The teams shared what rules were needed and what weren't. They advised how they could be enforced and took ownership of enforcing them as a team. They were specific about what backup was needed from management. That's shared responsibility in action.

Declutter your safety programme with Learning Teams

The key to decluttering your safety programme is co-designing improvements with the team. A Decluttering Learning Team is a great method for doing this as the team evolves its own safety framework. Perfect for ownership.

> It's important to identify 'the accumulation of safety
> procedures, documents, roles and activities that
> are performed in the name of safety, but do not
> contribute to the safety of operational work'.[14]

Let's explore the steps involved in a Decluttering Learning
Team.

1. **Contribution: the extent to which the activity has
    actual safety value.**

    Brainstorm where the clutter is in the organisation and
    choose the worst offender regarding safety processes
    that have become simply a tick-box exercise.

2. **Confidence: the certainty (either through evidence
    or strength of belief) with which this judgement is
    made.**

    Try to uncover evidence of the low efficacy of the safety
    activity. The example above included testimonies of how
    forms were used in practice and a systems analysis of the
    outputs. The collective experience of the Learning Team
    suffices as 'strength of belief' in action research, which
    is the scientific method used for gathering evidence in
    these situations.

3. **Consensus: the level of agreement about the safety value of the activity between those who mandate the activity and *those who are ostensibly kept safe by it.***

   This is where the Learning Team uncovers the issues and creates better solutions with the team who do the work. Getting them involved in the redesign will yield a new process that is far more effective.

Continuous evaluation and tweaking will be necessary to embed it after you've gone through this process and implemented a solution. You'll be amazed at team engagement and how they step up to take ownership of the new processes. It's a win-win. Let's look at an example.

I worked with a company that made their teams complete far too much tick-box paperwork. They couldn't finish it and get the job done within reasonable working hours. It meant the company was creating legal risk with partly completed forms.

Several processes overlapped in purpose, as the company had added to them over time. We know that duplication of safety processes is the enemy of engagement. When we used a Learning Team process to amalgamate and streamline the processes, the team became far more engaged and avoided legal risk.

Using Learning Teams to support your wellbeing strategy

I often get asked to help people with their wellbeing strategy. My recommendation is always that they should conduct a Learning Team and co-design their approach with their teams.

This makes perfect sense because, most of the time, the core contributor to wellbeing is work design. While mindfulness, positive thinking and other wellbeing initiatives are supportive, understanding how teams successfully navigate the conditions that challenge wellbeing will help you design appropriate interventions.

One organisation was keen to introduce a wellbeing approach during the pandemic years. Instead of a top-down, people and culture-led approach, we assembled a Learning Team with representatives from each site.

They grappled with the real issues affecting people's wellbeing, which mainly centred on production issues resulting from the extensive cost-cutting and redundancies due to the organisation's knee-jerk reaction to the instability of pandemic times. Ironically, business was better than ever between COVID-19 waves, and the team struggled with the lower headcount combined with the overall stress of ongoing uncertainty.

In this example, we see that the work factors significantly impact wellbeing. Using a model to unpack the aspects of work design to improve how work is conducted can deal with the systemic factors behind problems with mental health and stress in the workplace.

One organisation wanted to review a change management process they'd been through to hear the teams' perspectives on how it was conducted and learn more about how wellbeing could be maintained through inevitably stressful times. It turned out to be an insightful process. Management thought they were doing the right thing by painting all the possible scenarios, however, the teams said communications needed to be kept simple to avoid overwhelm. The outcome was an understanding that it was important to dial into where the teams were at and how they felt as change management processes were designed and communicated.

Supporting your learning framework with safety team support

Your safety team must shift to a coaching and facilitation role to support eliciting these insights throughout your learning framework. Dave Provan, Dave Woods, Drew Rae and Sidney Dekker wrote a paper *The Role of the Safety II Professional*, explaining that the safety role should shift from centralised control to guided adaptability.[15] Really, it is required reading for all safety professionals. You'll note the marrying of the language here with the primary purpose behind resilience engineering.

Evolving the New View of Safety contains a whole chapter on the role of the safety lead, in line with New View thinking. A full adaption of the role is required to support driving more self-direction into teams that will need far more coaching support than the current training our profession receives. For now, the material we covered in Chapter Eight on facilitating Learning Teams kicks off the principles behind the new role.

Learning Teams can (and should) be used for a wide variety of scenarios. Get creative. I've heard of and participated in Learning Teams to deal with IT issues, systems problems and the development of training packages. Their use is only limited by your imagination.

Now that you've broadened your toolkit and you're gathering intelligence from your team, it's time to transform the lens of your directors.

CHAPTER 12

Improving Safety Reporting and Assurance

Build a reporting approach that shows how teams are adapting to conditions and constraints

Key points

- Safety reporting and assurance traditionally focus on avoiding negatives rather than requirements for building capacity.
- Use current methods that are effective while adapting to contemporary methods. Make change iteratively rather than as a wholesale initiative.
- Recognise the political dynamics that influence what is and isn't reported to build trust and transparency.

At one large corporate, bonuses were linked to total recordable injury frequency rate (TRIFR) scores. The downstream effect on that company was a sorry tale.

One site confessed that a contractor had been seriously injured, and they'd tried to brush the accident under the rug. The team had shuffled the guy off to hospital, but then had a crisis of conscience and decided to report it after all.

The reality is, that everyone has to deal with an imperfect system as best they can. People are often left in the sticky situation of manipulating TRIFR data because of commercial pressures. We've all seen this type of thing happen, and honestly, no one wins from it in the long run. That's why I'm making a point of it.

The reality is that these situations sit in the context of ethical dilemmas, as we discussed in Chapter Six. Ethical dilemmas should be discussed at senior level in every organisation. It's an important topic to wrestle with, as there are no easy answers. This company referred to earlier relied on tenders for work, so TRIFR scores were crucial to making a good case for getting business. Again, it's the system, not the people. There's no judgement here, only a search for better solutions.

I don't know a safety lead who doesn't recognise that current reporting approaches aren't working that well. However, everyone is stuck on the way forward. If we take what we've got away, what do we replace it with?

As I finalise this book, there's a new lens on reporting in development. Hang on to your hats everyone, because change is coming. By the time you read this, you may well be heading down that track already.

The Better Governance report,[16] referenced in Chapter One, talks about a vision for developing governance capability that shifts us into:

- having the courage to admit when we don't have all the answers
- building the ability to better understand the worker's perspective on normal work
- creating the curiosity to seek insights into how work is prioritised, planned and resourced
- seeking a deeper understanding of the context in which the work is conducted
- developing the care needed to deliver the ethical responsibility inherent to the business activities.

The lynchpin is building transparency despite the inherent politicisation of information in organisations. We must shift the focus from qualitative metrics to quantitative insights to build a map of the organisation's collective intelligence. This supports our new definition of safety.

How do we create this new focus for our directors, leading them to ask much better questions? A radical new way of looking at safety-based reporting requires change agentry. To some extent, it must go through the courageous phase of testing readiness and running trials. Seriously reinventing your reporting approach requires a more mature culture and good traction with Safety II.

The international Safety Differently community is at the innovation stage of solving these problems. In 2021, Sidney Dekker, Kelvin Genn and Michael Tooma published the Due Diligence Index standard (DDI-S).[17] It is based on a paper published earlier by Dekker and Tooma and aligns with the six due diligence elements in Australasian law.[18] It uses all the concepts throughout this book for building capacity and resilience by turning them into suggested metrics with a resilience map. Your tools, developed in Chapter Eleven, should create the foundation.

The suggested output is in a style familiar to boards – a dashboard with report commentary. A scoring system could support the identification and visibility of resilience. The definitions set out in the DDI standard provide ideas for metrics which would map over to the dashboard.

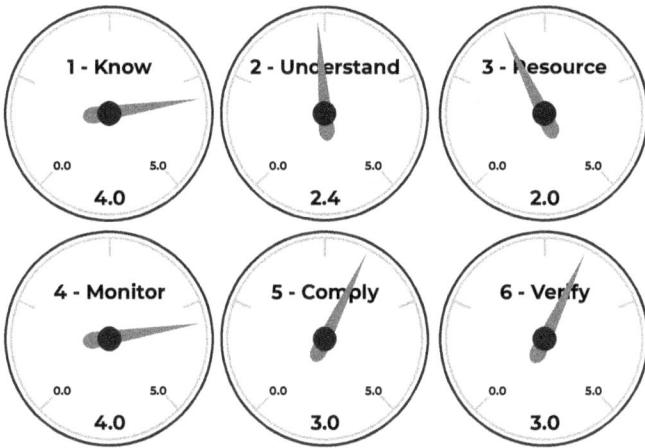

Rating and Descriptors

Rating	DDI-S Descriptor
1	Critical
2	Fragile
3	Brittle
4	Resilient
5	Optimal

Figure 26: Suggested resilience map and dashboard

For the rest of this chapter, we will look at how to take the fine ideas proposed in the DDI standard and use innovation to develop them further into something your organisation can use. After all, the intention of the standard was to create a starting point for everyone.

What is the problem with current assurance and reporting?

Current approaches to reporting on safety performance don't provide reliable data on the state of safety. Many of the commonly used indicators have no predictive value for future incidents and are prone to manipulation. TRIFR rates are still in frequent use, yet they are proven to correlate positively with catastrophic failure. The time spent gathering the data would be better spent out with the teams collecting intelligence.

Sidney Dekker says these metrics have a legacy in the twentieth-century tendency to measure everything in terms of productivity. It's a very insightful viewpoint. Unless we think boards of directors and CEOs are stupid, we must accept that they are measuring and understanding how much cost and downtime to the business occurs because of injuries. Given their mandate, that is fair.

However, they are measuring, as Sidney beautifully puts it, 'unsafety'. Think about it: measuring what causes people to be 'not safe' says little about what actually creates safety. The fact that directors could lose their homes or reputations still drives fear-based behaviour in designing safety management systems.

One of my clients expressed their frustration with directors asking questions that don't support discussion of the core issues driving safety. Another client bemoaned directors honing in on relatively minor issues, such as tiles out the front of a head office building. Directors are expected to ask questions, so we need to construct a new picture of what they should be looking at, to support them in asking us better questions.

The latest research shows (as mentioned in Chapter One) that incident rates are correlated to the presence of trust in organisations. Throughout this book, we've looked at stories which show that getting to the truth of your system's vulnerabilities can be hard but is well worth the effort for everyone. Transparency improves trust. When resourcing is supported by worker-led prioritisation, the team's ability to 'sense and respond' improves. So, we need to get a new set of data to the top table to enable this process.

Redefining the due diligence elements – shifting to a qualitative focus

Here's the rub. First and foremost, your new due diligence reporting approach requires a focus on qualitative insights in your reporting. These must give a robust map of capacity and its drivers, plus the current ability of teams to handle demands. Remember that we still report on the Safety I type indicators. As we've said all along – Safety II is an addition to current approaches.

When I talk about qualitative insights, I'm not talking about adding commentary (which everyone does), I'm describing changing the entire focus of your reporting to robust qualitative data. Let's look further at that.

You can't reduce the complex nature of how people work together to a number on a page. It's meaningless. Instead, use stories, videos, and general narratives to support real understanding. Even better, get your teams to show up at board meetings or write the reports in their words.

The insights should build an overall picture of constraints, conditions and what's helping and hindering team performance. That means directors need to adopt a mindset of curiosity and remain open-minded about the reality of the work the teams are dealing with. Rather than having all the solutions, they need to create a space for work teams to solve them along with the necessary resourcing.

I regularly see reports full of all the good things we're doing, all the initiatives and all the actions closed out. It's mostly the 'Looking Good Index', to quote Sidney Dekker. It's tantamount to window dressing and needs to change.

Here's a story that supports a focus on robust qualitative data.

A high-tech company I worked with (on staff) built what I'd call an early version of a Safety II-based reporting to leadership and directors based on the need for dynamic and real-time information on risks as they emerged.

Each week, during lean standups, the teams shared worker insights and actions were listed on boards that were visible to all. These were reviewed and monitored weekly by the team.

If anything wasn't actionable by the team, it was escalated for further input.

The safety owners compiled a filtered report based on risk-based insights the owners knew needed to be transparent to those further up the food chain. Where issues could be taken care of with cross-team committees, they were. Otherwise, they were highlighted to senior leadership in a monthly report.

No injury data was involved except safety alerts for incidents that everyone needed to learn from (again, qualitative). Directors received only these risk-based insights on a rolling basis. The work teams wrote all these reports, so my job as safety lead was simply to facilitate this process.

The secret sauce is, you guessed it, the use of AI. In a larger organisation, distilling the collective intelligence of the teams into a readable and digestible report for directors is achievable by harnessing the power of AI to draw themes and inferences out of large datasets.

It's all about the design of the dataset and teams understanding what they should be reporting to describe the state of team resilience. Trials are currently underway to build the AI code to collect and analyse the data. The principles are the same as using AI in a single Learning Team – just on a larger scale.

You'll want to play around with the keywords and prompts used to build a picture of capacities and demands. You'll need to focus on operational levers that help make work go smoother. Examples are prompts such as: 'training', 'parts', 'fuels', 'maintenance', 'materials'. You can ask about 'conditions and constraints' that are inherent in the work. Get busy with designing trials to test the output you get.

How, then, do you start building the necessary data to support your new definitions?

Process for introducing your due diligence framework

Few organisations have seriously jumped into building this framework at the time of writing due to the extent of genuine implementation of Safety II principles overall. Generally speaking, retaining your current reporting framework while building your new one is prudent.

You will develop your own approach to building your framework – always in context. Remember, however, here are the core steps to take:

Redefine safety: As an organisation, redefine safety from an absence of harm to the presence of capacity and resilience (or a similar definition that suits your organisation). This was covered in earlier chapters. The concepts need a clear introduction with your directors. Engaging them will ensure they support innovation in adapting your reporting processes. Start small and build from experiments, trials and pilots.

Define due diligence elements: Redefine each due diligence element to support your definition. We'll look at a possible worked trial in Table 6.

Review current metrics: Determine which of your current metrics meet your definitions so you know what to keep. Avoid wholesale reinvention by retaining what you have and adding to it.

Trial new tools and metrics: Fill the gaps by trialling qualitative information and new metrics using your Agile Learning Framework. Use AI to analyse themes in the data you are collecting. This will enable output that understands how to support your teams better.

Practical reporting

Let's consider some ideas for building your Safety II reporting in practice. Remember that, as with all Safety II approaches, you will want to try new ideas iteratively and use experiments and trials to get started.

Most organisations embarking on this journey start with 'leader walks', where senior executives and directors have a KPI to get out into the field and have regular safety conversations with the team. Coaching your leaders in new methods of enquiry with better questions gets them accustomed to the kind of intel elicited and sets the stage for more robust data gathering.

With this gathered intelligence, including analysis across all the data, themes can be drawn out. Combined with insights gathered through other tools and Learning Team reports, your organisation will be well on its way to viewing safety differently.

However, ultimately, organisations need a line of sight to where they're heading concerning reporting and data to analyse. Don't stop short of envisioning overhauling your reporting output over time. The following table (Table 6) suggests definitions with associated qualitative data.

	Safety II definition	Qualitative data
Know	Filtered worker insights based on capacity and demands (prioritised by workers).	Reported qualitative observations/insights analysed by AI for capacity and demands themes that need resolving.
Understand	Work as done vs work as imagined – key weak signals.	Analyse Learning Team reports for the month to highlight weak signals especially relating to critical risks and controls. Use AI to support the process.
Resource	Gap between current resourcing and required resourcing for Know and Understand.	Analyse the cost required to sort the gaps in the two previous sections.
Monitor	Number of learning hours per worker.	This metric could be given based on manpower hours similar to TRIFR.
Comply	Any outstanding compliance issues.	Report your usual compliance findings based on Safety I style audits, inspections, and other necessary compliance activities.
Verify	Sampled feedback from workers on whether the Safety II system is being used well and whether resolution is happening.	Conduct a rolling feedback loop with a three-month follow-up to Learning Teams to ensure workers' insights are heard and acted upon.

Table 6: Definitions with associated qualitative data

Given the current state of AI (at the time of writing), you may need to consider building bespoke AI capability to handle the scale of data your organisation collects. Consider analysing data in your safety and operational management systems. After all, capacity and resilience sit within how the team operates, not in the safety system which attempts to control work.

Now that you have some new ideas about your reporting, you will hopefully see how this links back to where we began at the start of this book. We wanted to find new ways to humanise the systems in our businesses in a win-win manner.

We looked at how to build resilience in our teams to get safer outcomes through work going smoothly and well. We challenged some of the notions of how safety is viewed in the workplace.

Understanding the system vulnerabilities by focusing on what creates success in teams is quite a flip in thinking and will take time to drive through your business. You'll want to start small and test the ideas. Demonstrate the benefits to leadership and gain co-option of the processes as you go.

We've examined an entire Learning Teams framework, right through to using AI to revamp reporting. These system changes will create a new lens and focus for your directors and leaders. Ultimately, your teams will be seen as the smart, capable people they are when supported and resourced to handle variability better. The promise of high-performing teams makes the effort worth it.

Epilogue

As we shift the focus from compliance with reductionist systems and structures to resourcing our teams well and providing the support they need to remain resilient, we can create a high-performance culture that benefits everyone.

While this is the end of this book, it is really just the beginning. New View Safety is, quite literally, new, being only ten years old in practical terms. There is much territory to discover. The path of innovation has just begun.

You'll want to start thinking about ideas for the future of the Safety Differently movement. That's what my next book is about.

While starting with Learning Teams prepares the way and sets the foundation, if you think you've got this licked, trust me, there is far more to build. As much as the concepts and ideas of Learning Teams have become popular and accepted, they also have many challenges. They are time-consuming, resolve matters discretely, and still enable the current hierarchical model of working together. However, they are a necessary stepping stone to creating the trust needed for the next stages.

As the table at the start of this book showed, we have covered the three basic concepts needed to prepare your organisation to embrace New View Safety. We've run through how to introduce the new ideas to your organisation. There are new ways of looking at time-honoured problems that provide early benefits and gains and demonstrate your leadership.

Maturity	Stage	Focus
Beginner	Introduction to concepts	Getting ready
Elementary	Readying the organisation	Introducing new ideas
Intermediate	Building a foundation	Learning framework
Competent	Establishing new practices	Co-design with teams
Advanced	Team-managed safety	Self-directed work teams
Evolved	New world approaches	Co-governance, ethics and transparency

Table 7: Looking towards evolving your system over the coming decade

The next stages are where the real transformation begins. We look at the systems and structures that need adapting and evolving. We start moving from listening and building trust to enabling team ownership and even handing over authority for safety management to our teams. That may seem ambitious now, but look at what we've achieved over the last decade. We're only limited by our collective imagination.

Before we finish, let's take a peek at the next book to whet your appetite.

Co-designing your safety programme with your team

Have you come across the concept of anti-fragility? It means the more you stress a system, the stronger it becomes. What we try to do from here is harness the power of anti-fragility.

Our organisations think letting go of control will reduce system efficacy. However, paradoxically, it has the opposite effect when best practice is followed (and there are certainly some very important ingredients to make it work).

People become more responsible and accountable and contribute more. Problems are transparent and easily solved. Fixes are more efficient. Teams are enthusiastic and engaged with safety. They are forthcoming and willing to participate. I've seen this personally. The solutions to age-old problems in our profession are right under our noses. Ultimately, we give our people the ability to 'sense and respond', which means we harness their full potential in our organisations. There are some preconditions to make this work, however.

Your teams are entirely capable of working with you to make most of your safety decisions. Work with them to decide on:

- safety rules
- safety procedures
- documentation needs
- training needs
- critical risk continuous improvement

- monitoring and checking
- system design
- safety programme design.

Before handing full authority to your teams, go through a transitionary phase by allowing them to assume as much responsibility as possible for their own safety programme that is more integrated with work processes. That ultimately means reducing reliance on a separate safety management system monitored by safety leads.

This could entail the following types of activities:

- instigate work team lead prestarts, toolboxes and inductions
- integrate operational work review activities with resilience engineering
- co-design training programmes
- reduce paperwork clutter that isn't necessary for the teams to do a good job
- support a full Freedom within a Framework approach
- give workers input into budget and capex prioritisation.

These processes will make more sense when you join me in my next book. We will walk through how to build your team's capability by adopting Freedom within a Framework, decluttering, self-directed training approaches, and innovative ways of resourcing teams.

Beyond this (and yes, there is more), we'll take this a step further and hand authority for safety over to the team themselves. This is advanced and requires some systems redesign and cultural development to support high-trust teamwork.

Understanding self-directed work teams

Given current paradigms and structures where authority for safety rests with a safety team, it can be hard to imagine running safety via self-directed work teams. The radical element is when the safety team hands that responsibility to the teams. The safety lead then transitions to a performance coach role.

Yet this is the way of the future as the younger generations who value these approaches become increasingly influential in running our organisations. I set up a framework using these approaches several years back, and the high-tech organisation I mentioned in Chapter One is still using it effectively to support team performance and agility. All with the approval of our regulator. It just requires thinking through how team safety needs will be met, including procurement, budget approval and technical support.

We can adapt our safety systems using the wisdom from literature on self-managing teams. Frederic Laloux, the author of the popular book *Reinventing Organisations,* describes the processes involved with self-managing teams as follows:[19]

- team-based decision-making
- granular roles

- information transparency
- performance management
- conflict resolution.

Management theorist Brian Robertson created a framework and methodology for team management that we can learn from (Figure 27). In his book *Holacracy,* he shows how to shift away from a hierarchical way of working together.[20] Dynamic roles support adaptability to create resilient teams that can 'sense and respond' effectively over time, meaning they can keep themselves safe with less dependence on safety management systems.

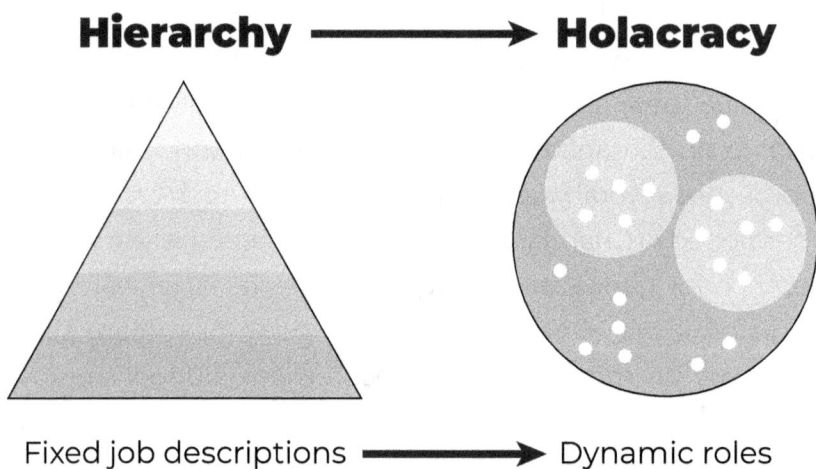

Hierarchy ⟶ Holacracy

Fixed job descriptions ⟶ Dynamic roles

Figure 27: From hierarchy to holacracy

To understand how this works in practice, we must unpack how to create a team-managed safety structure. I will demonstrate how to move further towards an eventual safety programme that is fully owned by the team without the need for a safety management system.

It will mean creating lean systems and processes to support their needs. We look at how to do this in a way that still meets the regulators' requirements and gains approval with safety cases. The two core ingredients are a high standard of ethics and full transparency with any regulators, which we'll explore.

From there, you will need to evolve your teams further to support the trust and ownership required to make team-managed safety work.

Running team-managed safety

High-performing teams are central to team-managed safety, as it is based on an understanding of intrinsic motivation. Given your perception of your team's current performance, you might think this isn't possible in your organisation. However, the self-directed teams I've worked with confirm that an upward spiral begins when authority, autonomy and agency are given. The teams demonstrate an increase in responsibility and ownership. Some organisations have teams already operating like this, so find out the secrets to their success and learn from them. Remarkable things can be achieved.

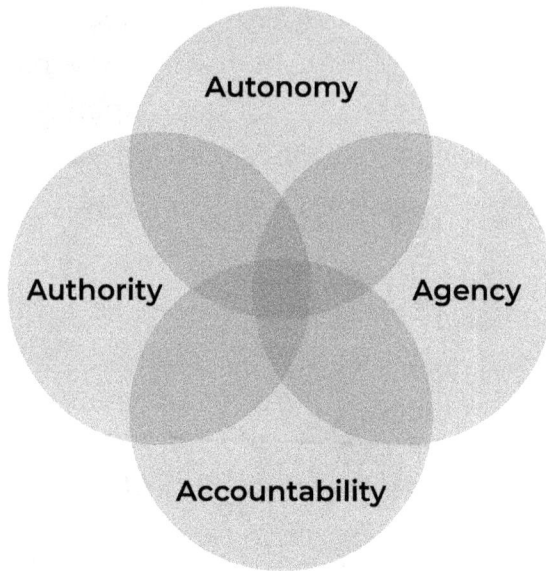

Figure 28: Elements of empowerment

Creating the space to support the team requires 'host leadership'. In short, managers learn to let go of control and give authority to the team to make decisions. Giving teams agency requires peer-to-peer accountability to work. New Zealand's All Blacks rugby team uses these approaches to support their world-leading performance. *Evolving the New View of Safety* will explore case studies and how to implement this.

We'll also examine the necessary mental and team skills required to make it work. We'll discuss the team's role and needs and how to ensure they get the necessary support. The team will need to go through a development process to reach the maturity to manage their safety. Figure 29 shows the changing role of the supervisor from one who holds team activities together, to a leader who lets the team run things and is there for support.

EVOLUTION OF A TEAM LEADER'S ROLE

Stage 1
Start-up team

Authority
Expert
Teacher
Problem solver
Coordinator
Team supervisor
Mentor

Stage 2
Transitional team

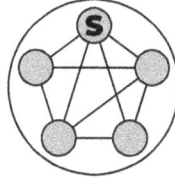

Shared authority
Monitor
Helper
Example setter
Evaluator
Information provider
Link to other teams

Stage 3
Well-trained,
experienced team

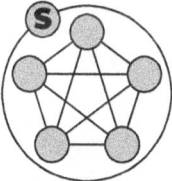

Boundary manager
Auditor
Expert
Resource provider
Goal setting guider
Information provider
Protector/buffer

Stage 4
Well-trained,
mature team

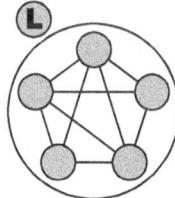

Boundary leader
Shared values
Coach
Champion
Counsellor
Resource provider
Supporter
Shared responsibilities

S supervisor or manager **L** Leader

Figure 29: Developmental stages of team-managed safety
– the shifting role of supervisor to supportive leader

Your organisation is likely already developing around these principles, so you'll find plenty of embedded capability. As in this book, we start with trials and iterate our way through change.

The Safety Differently movement has just begun and there is much territory to discover. The way ahead needs innovating into being. I hope you'll join me.

Do send me a LinkedIn request. I post news and articles regularly and will let everyone know when *Evolving the New View of Safety* is available.

If you want to see my latest work, pop along to my website **www.monihogg.com**.

If you are having a go at self-managing teams, I'd love to hear your story. Drop me a line if you fancy a Zoom coffee; I'd love to hear what you're up to.

About the Author

Moni Hogg is a senior health and safety practitioner specialising in the Safety Differently/Safety II approach. Moni was the first workplace safety manager for Rocket Lab USA. She was a finalist in the 2019 HRINZ Health, Safety & Wellbeing Award for success with contemporary safety design and innovation in aerospace.

Moni is nationally recognised in New Zealand as a pioneer in contemporary safety approaches. One of her major clients won the 2023 Safeguard Safety II Award for successfully shifting to New View Safety leadership in the challenging civil construction sector.

As an innovator and early adopter of Safety Differently, Moni started her first trials in 2015. She has extensive experience mentoring teams and organisations as they start or deepen their implementation of Safety II right across the industrial spectrum in New Zealand. Moni is a highly experienced trainer, facilitator and coach. She has collaborated on projects with Art of Work (an Australian-based consultancy) and the Worksafe NZ Innovation team, who have focused on using contemporary safety approaches in the regulatory context.

Moni has delivered two national roadshow speaking tours around the regions of New Zealand for the NZ Institute of Safety Management, teaching a roadmap for implementing New View Safety. She hosts a regular webinar series for Women in Safety Excellence showcasing case studies of successful Safety Differently implementation. Moni speaks internationally on podcasts and delivers a masterclass on self-managing teams.

With more than twenty years of experience in safety, after ten years in construction operations, Moni has consulted and trained across most New Zealand industries, including construction, manufacturing, farming, health, transport, engineering, emergency services and government organisations. This experience gave her broad knowledge of the realities of work across most sectors.

Moni spent her twenties working in housing teams as a construction supervisor for New Zealand's leading construction conglomerate, Fletcher Building. In the late 1990s, she broke ground as a young female in construction management. Ultimately, this work set her up for success with New View Safety because she has first-hand experience as an operations lead and understands the day-to-day challenges of creating Better Work in teams.

She has competed in entry-level motorsport, loves hiking in the national parks across New Zealand and lives in central Auckland.

For more information, visit **www.monihogg.com**.

References

1. Long, R. (2015). *The ethics of safety*. Available from: https://safetyrisk.net/the-ethics-of-safety/

2. New Zealand Business Leaders' Health & Safety Forum. (2023). *Findings and recommended actions to improve health and safety governance in Aotearoa New Zealand*. The Better Governance Report Series. Available from: https://www.forum.org.nz/assets/Summary-report.pdf

3. Bitar, F.K., Chadwick-Jones, D., Lawrie, M., Nazaruk, M., & Boodhai, C. (2018). Empirical validation of operating discipline as a leading indicator of safety outputs and plant performance. *Safety Science*, Volume 104, April 2018, pages 144-156. doi:https://doi.org/10.1016/j.ssci.2017.12.036.

4. Hollnagel, E. (2013). A Tale of Two Safeties. *Nuclear Safety and Simulation*, Vol 4, Number 1, March 2013.

5. Dekker, S.W.A. (2011). *Drift to Failure: From hunting broken components to understanding complex systems*. CRC Press.

6. Campbell, J. (2014). *A Hero's Journey*. New World Library.

7. Bushe, G & Marshak, R. (2015). *Dialogic Organisation Development: The theory and practice of transformational change*. Berrett-Koehler Publishers Inc.

8. Rogers, Everett. (2003). *Diffusion of innovations* (5th ed.). New York, NY: Free Press.

9. Long, R. (2019). *The Social Psychology of Risk Handbook*. Scotoma Press.

10. Dekker, S.W.A. (2017). *Field Guide to Understanding Human Error*, 3rd edition, CRC Press.

11. Dekker, S.W.A. (2017). *Field Guide to Understanding Human Error*, 3rd edition, CRC Press.

12. Dekker, S.W.A. (2007). *Just Culture: Balancing safety and accountability*, Ashgate Publishing Ltd.

13. International Council on Mining and Metals. (2015). *Health and Safety Criticial Control Management*. Available from: https://www.icmm.com/en-gb/guidance/health-safety/2015/ccm-good-practice-guide

14. Rae, A.J., Provan, D.J., Weber D.E. & Dekker, S.W.A. (2018). Safety Clutter: the accumulation and persistence of 'safety' work that does not contribute to operational safety. *Policy and Practice in Health and Safety*. August 2018. https://doi.org/10.1080/14773996.2018.1491147

15. Provan D., Woods, D., Dekker, S. & Rae, A. (2019). Safety II professionals: How resilience engineering can transform safety practice. *Reliability Engineering and System Safety* 195 106740.

16. New Zealand Business Leaders' Health & Safety Forum. (2023). *Findings and recommended actions to improve health and safety governance in Aotearoa New Zealand*. The Better Governance Report Series. Available from: https://www.forum.org.nz/assets/Summary-report.pdf

17. Due Diligence Index Council. (2021). *Due Diligence Index – Safety Standard*. https://www.duediligenceindex.online/

18. Dekker, S. & Tooma, M. (2021). *A capacity index to replace flawed incident-based metrics for worker safety.* Griffith University.

19. Laloux, F. (2014). *Reinventing Organisations: A guide to creating organisations inspired by the next stage of human consciousness.* Nelson Parker.

20. Robertson, B. (2015). *Holacracy: The New Management System for a Rapidly Changing World.* Henry Holt & Co.